101 Questions & Answers About **CB Antennas**

by

Jim Ashe

Edited by

Leo G. Sands

HOWARD W. SAMS & CO., INC.
THE BOBBS-MERRILL CO., INC.
INDIANAPOLIS · KANSAS CITY · NEW YORK

FIRST EDITION

FIRST PRINTING—1971

Copyright © 1971 by Howard W. Sams & Co., Inc., Indianapolis, Indiana 46206. Printed in the United States of America.

All rights reserved. Reproduction or use, without express permission, of editorial or pictorial content, in any manner, is prohibited. No patent liability is assumed with respect to the use of the information contained herein. While every precaution has been taken in the preparation of this book, the publisher assumes no responsibility for errors or omissions. Neither is any liability assumed for damages resulting from the use of the information contained herein.

International Standard Book Number: 0-672-20749-4
Library of Congress Catalog Card Number: 73-149394

TK 6565
A6
A7

Preface

A survey of readers of *CB Magazine* revealed that 86 percent of the readers responding wanted to know more about antennas. CB (Citizens band) radio equipment dealers report that antenna sales are booming. Mnst CB'ers (Citizens band operators) are anxious to get as much range as possible. Since transmitter power is limited by FCC rules and regulations, the only lawful way to increase transmitting range is to use a more efficient antenna in the most effective way. Which antenna to buy and how to install it are the two most often asked questions. But, there are no specific answers that apply to every situation because of the variables.

This book is for the practical CB'er as well as the CB equipment salesman and service technician. It is not a theoretical treatise on antenna design. It does, instead, present the answers to the 101 most often asked questions about CB antennas.

JIM ASHE

Contents

PART 1

Your Antenna Is Very Important

1. Basically, what do antennas do? 10
2. What is the theoretical maximum range of a CB rig? 11
3. Are elaborate antennas really better than simple ones? . . . 12

PART 2

Basic Antenna Concepts

4. What is a field? 14
5. What are electric and magnetic lines of force? 15
6. What is a dipole? 16
7. How do antennas radiate? 16
8. What are dipole and folded-dipole antennas? 17
9. What are Hertz and Marconi antennas? 18
10. What is antenna polarization? 20
11. What is antenna radiation resistance? 21
12. What is a ground plane? 21
13. What is a loaded antenna? 22
14. What is antenna gain? 23
15. How is antenna gain achieved? 24
16. What is antenna beam width? 24
17. What is antenna capture area? 25

PART 3

Antenna and Site Performance

18. How do radio waves travel from one place to another? 28
19. What is the ionosphere? 29
20. What is site noise? 30
21. What is a decibel? 31
22. What is an "S" unit? 34
23. How does effective antenna elevation affect range? 35
24. When should vertical polarization be used? 36
25. When should horizontal polarization be used? 37
26. What is a passive repeater? 38
27. Is it lawful to use a gain antenna? 39
28. How can two or more antennas be used? 39
29. Can the same antenna be used for two or more transceivers? . . 40
30. What is an antenna amplifier? 41
31. Can a CB transceiver be mounted near the antenna
    but controlled from a distance? 42
32. How can a range field survey be conducted? 42
33. Is an earth ground required? 43
34. How can skip reception and transmission be prevented? ... 44

PART 4

Real Antennas

35. What is a tradeoff? 46
36. What is antenna bandwidth? 49
37. What is SWR? 49
38. What is a Yagi antenna? 50
39. What is a dual polarized antenna? 50
40. What is a quad antenna? 52
41. What is a "low-profile" antenna? 53
42. What is a "disguise" antenna? 53
43. What is a "scanner-type" antenna? 53
44. What is a fiberglass antenna? 54
45. Can an indoor antenna be used for CB? 54
46. Can a long-wire antenna be used for CB? 55

PART 5

Feeding Antennas

47. What is a transmission line? 58
48. What is "mismatching?" 60
49. Why is a coaxial cable nearly always used
    as a transmission line? 61

50. How do coaxial cable types vary? 61
51. What kind of coaxial cable should be used? 62
52. How should coaxial cable be spliced? 63
53. How are connectors attached to coaxial cable? 64
54. What is a solderless coaxial cable connector? 64
55. How is a CB transceiver tuned to the antenna system? . . . 64
56. What is a dummy antenna? 64
57. What effect does SWR have on transmission line losses? . . . 66
58. How can coaxial cable be tested? 68

PART 6

Installing Base-Station Antennas

59. What is the difference between antenna height and
    effective antenna elevation? 72
60. How is the legal 20-foot height limit applied? 72
61. How can TVI be eliminated or reduced? 73
62. What is an antenna matcher? 75
63. Is a lightning arrestor required? 75
64. What are some basic precautions that can be taken
    against lightning damage? 77
65. How should a CB antenna be mounted with respect
    to a TV antenna? 77
66. Can a CB antenna be used with both a CB transceiver
    and a 30-50 MHz band monitor receiver? 78
67. Can a TV antenna rotator be used with a
    beam-type CB antenna? 78
68. What kind of antenna can be used on an apartment terrace? . . 79
69. Can a CB antenna be mounted in an attic? 79
70. How can a CB antenna be mounted on a water tower? . . . 80
71. How can coaxial cable be run from
    outside into a building? 80
72. How can electrical connections be kept from corroding? . . . 81
73. Why should a tower be accurately vertical? 81
74. How should a tower be guyed? 82

PART 7

Installing Antennas on Motor Vehicles and Boats

75. What is the best place on a car for mounting an antenna? . . 84
76. What mistake in a truck or van antenna installation
    leads to poor results? 85
77. Can the same antenna be used for CB and an auto radio? . . 85
78. Where is the loading coil located in an
    electrically shortened antenna? 86
79. What is capacitive loading? 87

80. What are the sources of objectionable mobile noise? 87
81. How can an external antenna be connected to a walkie-talkie? . 88
82. Is it necessary to ground an antenna to a car body? 89
83. What kind of antenna can be used on a convertible car? . . . 89
84. What is a magnetic mount antenna? 89
85. What is a clip-on antenna? 90
86. What is a side-mount antenna? 90
87. What is a spring mount? 91
88. How can an antenna be mounted on a car
without drilling holes? 91
89. Is there such a thing as a mobile beam antenna? 92
90. What kind of antenna should be used on a boat? 93
91. Can a wire antenna be installed on a boat? 93

PART 8

Testing Antennas

92. What preliminary check may make electrical
tests unnecessary? 96
93. Why is a CB receiver a poor test instrument? 97
94. How can antenna radiation be monitored? 97
95. What is a comparison test? 98
96. What is an attenuator? 99
97. Why is reflected power important? 100
98. How is reflected power measured? 101
99. How is "VSWR" misleading? 102
100. What can be learned by measurements
at the antenna terminals? 102
101. What transmission line lengths are preferable for test work? . 103

Part 1

Your Antenna Is Very Important

The antenna is a vital part of any radio communication system. At the transmitting end of the radio link, it couples the transmitter into the great electric circuit that is all space. At the receiving end, it couples a tiny remaining bit of the transmitted signal back into wire circuits again, and into the receiver.

If the radio path from transmitting to receiving antenna has very good properties, antenna efficiency is unimportant—until some emergency arises or it is necessary to work over some other poorer path. For this reason, any antenna installation worth putting up at all is worth putting up right.

Antenna installation cannot be done casually. Space transmission losses due to distance and intervening obstacles are highly variable. But the installation designer can choose an arrangement that most effectively uses the potentialities of the site. This is entirely up to the designer, who may be taking on a considerable responsibility. Nothing else in the entire system, provided the transceiver is in good shape, is as important for good results as the antenna installation.

1

Basically, what do antennas do?

Antennas provide means for feeding radio signals out of a transmitter and into a receiver. Communication effectiveness depends strongly upon antenna design, since various antennas at a particular site may have the same effect as changing the transmitter power by a factor of 10 or more.

The competent antenna engineer is familiar with three key antenna facts. He knows an effective antenna occupies a fair amount of space. He understands the same basic principles apply to both receiving and transmitting antennas. And, he comprehends how radio signals in space travel in straight lines complicated by interactions with the earth, with absorbing and reflecting surfaces on it, and with the air and ionosphere (Fig. 1-1).

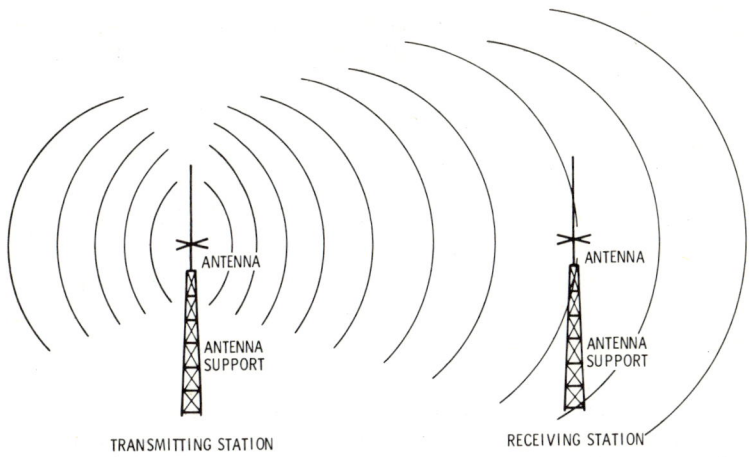

Fig. 1-1. Only a small part of the transmitted signal is captured by the receiving antenna.

2

What is the theoretical maximum range of a CB rig?

There is no sharply defined maximum range. Rather, beyond line-of-sight transmission is the usual performance. Part 95 of the FCC Rules and Regulations fixes a 150-mile limit. But, radio amateurs sometimes communicate around the world by using 5-watt and smaller transmitters operating on frequencies near the Citizens band.

Under certain conditions, low-power transmitters can be heard over tremendous distances. A CB rig is adequate for good communication with another CBer on the moon, given a fair antenna and a quiet band. The output of the Mariner 4 transmitter that returned the first close-up Mars pictures was 10 watts. It was later heard on the other side of the sun, 191 million miles away, still reporting on the Mariner's experiences in space.

3

Are elaborate antennas really better than simple ones?

Often, yes. If a simple antenna isn't enough, the controlled properties of a more elaborate antenna are indicated. A margin for efficiency loss due to aging and unexpected needs should be considered when choosing an antenna. Here are two reasons for choosing a gain antenna:

Interference—A beam antenna has directional selectivity (Fig. 1-2). This reduces the bad effects of band congestion and noise interference. Correct orientation of a beam antenna may reduce a persistent unwanted signal by 100 times or more.

Weak Signal—If a beam antenna is placed at one end of a communications link, its effect is that of increasing effective radiated power by its own gain. A beam antenna at both ends is still more effective.

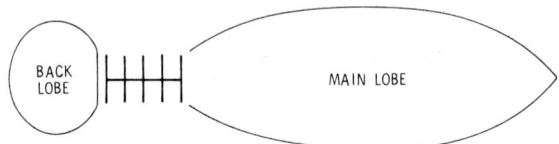

Fig. 1-2. A beam antenna provides gain by concentrating the radiation and reception into a narrow beam.

Part 2

Basic Antenna Concepts

What really happens when your antenna radiates a signal into space, or acquires some small part of another signal? The answer begins, as so many useful facts do, in history.

Men have known something of electric charges and fields, of magnets and magnetic fields, for at least 2500 years. The first book based on anything resembling electrical laboratory study was published in 1600 A.D. About 1750, Benjamin Franklin guessed that electricity and magnetism were somehow related, because he knew that the compass of a ship could be reversed if the ship were struck by lightning. Franklin also originated the terms "positive" and "negative" to replace many curious terms used in his day.

The connection between electricity and magnetism cleared up in 1865, when James Maxwell worked out his equations that are the basis of modern radio. He showed how wires could radiate electric energy, and that this energy would travel through space at the speed of light.

By 1900, electric machinery and lighting equipment were in factory production. Thompson had puzzled over some little bits that carried an electric charge, and Johnstone Stoney had named them "electrons." Hertz and Marconi were developing Maxwell's work into long-distance communications systems. DeForest and Fleming were beginning the work that led to the triode vacuum tube. By then, the basics in this chapter were well understood.

4

What is a field?

In radio there are three kinds of fields: a fixed electric field, a fixed magnetic field, and a moving electromagnetic or radio field. Since these fields cannot be detected by any human senses, they are defined as being regions in space where their presence produces some observable result.

Electric Field—A region in space where an electric force acts upon an electrically charged particle. The object may be a piece of dust, a cat's hairs bristling, a capacitor plate, or an electron in a cathode-ray tube, for instance. An electric field in a current-carrying conductor is responsible for the electron motion along the wire.

The direction of an electric field is said to be the direction a positive electric charge wants to move in it. This convention began the dispute about which way current flows. Electrons are negative, and flow against the conventional field direction.

Magnetic Field—A region in space where a magnetic force is exerted on a magnetic pole. The direction of the field is conventionally the direction a North-seeking pole wants to move. The magnetic field of the earth, for instance, goes from its South pole to its North pole.

Radio Field—A region in space where some detector, such as a field-strength meter or a radio receiver through its antenna, detects the presence of radio frequency energy. A radio field is a joint electric and magnetic field moving at the speed of light.

5

What are electric and magnetic lines of force?

An electric "line of force" is a line, in imagination or on paper, representing the course a positive test charge would follow if released in the field. Electric lines of force don't really exist, since electric fields have no detectable structure or boundaries.

Such a vagueness is very difficult to get a mental grip on. Around 1830, Michael Faraday came up with the "lines of force" convention. He talked as if electric fields were composed of lines and tubes of force. They aren't, and Faraday knew it. But his approach works so naturally and well, many people think, even today, that the lines of force are really there.

Magnetic "lines of force" show how a North-seeking magnetic pole would move in a magnetic field. Iron particles falling onto a piece of paper with a magnet under it seem to arrange themselves on the lines of force. But repeating the experiment produces a picture different in detail, though recognizably similar.

Electric and magnetic lines of force only suggest the shape of a field. They help us picture it and think about it; that's all.

6

What is a dipole?

The original dipole is very different from the radio antenna that goes by the same name. In engineering and physics, the idea of two equal but oppositely charged poles gets much use. This charge arrangement is naturally called a "di-pole." The physicist's dipole helps explain how a real dipole antenna works, and why it radiates in a doughnut pattern instead of equally in all directions, as shown in Fig. 2-1.

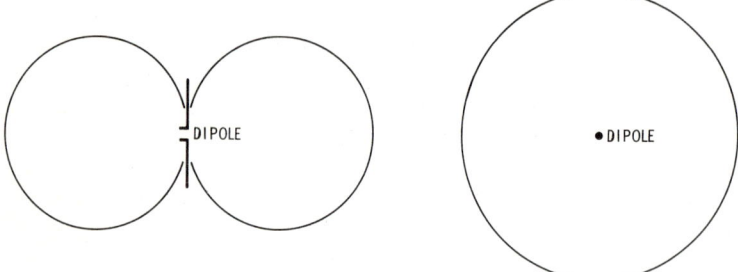

(A) Pattern is bidirectional with the dipole horizontal to the earth.

(B) Pattern is omnidirectional with the dipole vertical to the earth.

Fig. 2-1. Radiation patterns of a dipole.

How do antennas radiate?

Antennas radiate by feeding an electric current into space. Any scheme that puts an electric current through space, or through a conductor in space, radiates electric power. This was a surprising conclusion when first worked out by James Maxwell in 1865. Now we know it's simply an application of the fact that an alternating current will flow into one lead of a capacitor, through space between the capacitor plates, and out the other lead.

The basic radiating system is the physicist's dipole. If an alternating voltage is fed to this "di-pole," an alternating electric field appears, and a current flows through the space. This sets up an electromagnetic field that spreads outward at the speed of light. A radio signal is radiated.

8

What are dipole and folded-dipole antennas?

A dipole antenna is the radio engineering form of the physicist's "di-pole" electric charge arrangement. Instead of two metal spheres, it is a pair of equally long conductors placed end-to-end, as shown in Fig. 2-2. Radio frequency power is fed to the center terminals of the conductors, and flows as if into a 72-ohm resistor. The power is converted into radio waves, rather than heat. At resonance, a dipole antenna is slightly less than a free-space wavelength long.

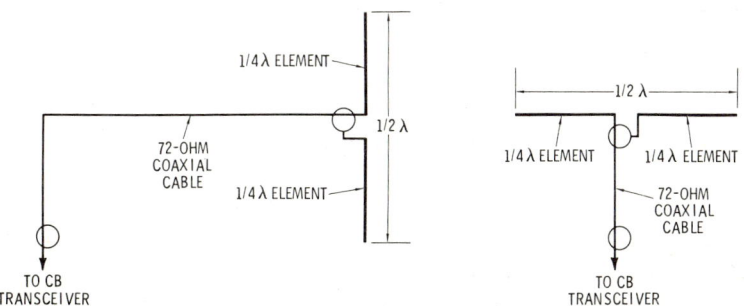

(A) Vertically polarized dipole.   (B) Horizontally polarized dipole.

Fig. 2-2. Dipole antenna systems.

A simple change does not affect the way the dipole works, but increases its input resistance to 300 ohms. The radiating conductor is doubled and joined at the ends as shown in Fig. 2-3. To feed the same amount of power into this antenna as into a basic dipole requires double the voltage and half the current. Its input resistance is four times 72 ohms, or practically 300 ohms.

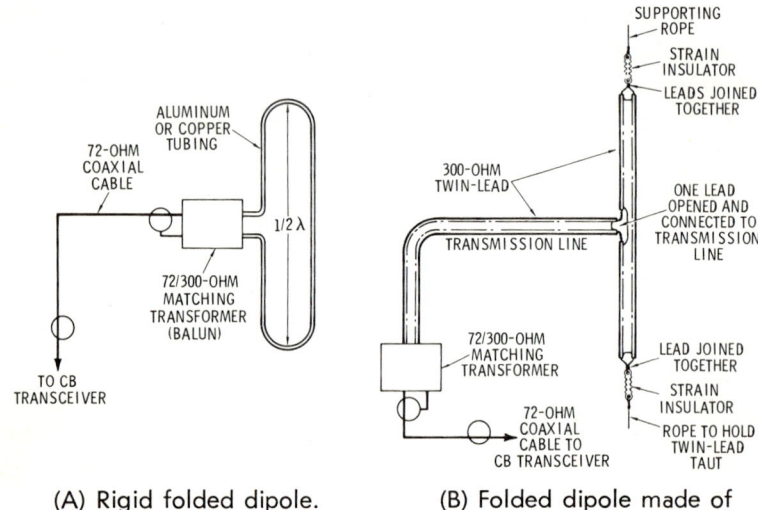

(A) Rigid folded dipole.

(B) Folded dipole made of twin-lead wire.

Fig. 2-3. Folded dipole antenna systems.

9

What are Hertz and Marconi antennas?

Both are basically dipole antennas. The basic dipole is composed of two equal parts and is fed a large current at its center. It isn't necessary for the dipole to be balanced, or current-fed. The Hertz and Marconi antennas hide their dipole ancestry, but working out their electric and magnetic fields shows they are still dipoles.

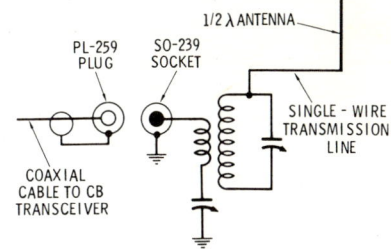

Fig. 2-4. Example of a tuning circuit for an end-fed half-wave antenna.

Hertz Antenna—At resonance, it doesn't matter if a half-wave gets a high current at its center or a high voltage at its end. The Hertz antenna is a voltage-fed half-wave dipole. Voltage feeding is very convenient in certain difficult sites since the Hertz can be bent into available space and then trimmed or loaded to resonance.

Usually, a Hertz antenna is fed by a tuned circuit, as shown in Fig. 2-4. Radio power is link coupled to the circuit through a length of coaxial cable. Because the current is low at this high-voltage feed point, ground resistance losses are a small factor in installation design, as compared with the Marconi antenna.

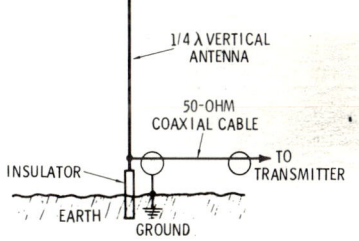

Fig. 2-5. A Marconi antenna.

Marconi Antenna—If a one-foot ruler is placed square to a a mirror, the casual observer thinks he sees a two-foot ruler. It's hard to see where the ruler leaves off and its reflection begins. This works at radio frequencies. The Marconi antenna is one-half of a dipole, only a quarter-wave long (Fig. 2-5). Its other half is a radio reflection in a metal surface that is often called a *ground plane*. The reflecting surface can be reduced to a few quarter-wavelength radial conductors, with only minor antenna efficiency reduction, to form what is commonly known as a *ground plane antenna* (Fig. 2-6).

A Marconi antenna is usually fed through 50-ohm coaxial cable, although its input resistance is below this. The resulting mismatch does not greatly reduce efficiency.

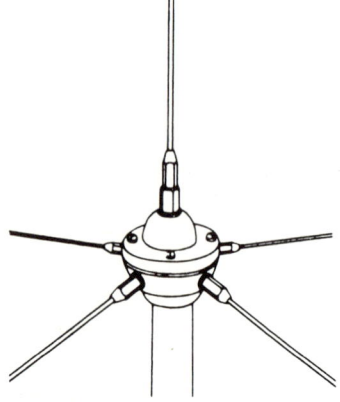

Fig. 2-6. A ground plane antenna.

10

What is antenna polarization?

The polarization of an antenna is its electric field orientation. If the electric field is horizontal during a voltage maximum, the antenna is said to be horizontally polarized. A horizontal dipole radiates a horizontally polarized field at right angles to its axis. Off the ends and somewhat away from the axis (where the radio field is zero in free space), the field of the dipole is vertically polarized.

Antenna polarization isn't always obvious. A vertically mounted cylindrical antenna may be chosen for its horizontal gain and nondirectional performance. At first sight, it looks vertically polarized; it is really horizontally polarized.

Vertical polarization is almost universally used in base-to-mobile and mobile-to-mobile systems. Horizontally polarized beam antennas are often used when only fixed point-to-point communication is required.

11

What is antenna radiation resistance?

The radiation resistance of an antenna is the ratio of voltage to current at which the antenna accepts power. This is measured at its terminals. It turns out that if an antenna is replaced by a resistor equal to its radiation resistance, the driving circuit cannot tell the difference. The resistor dissipates the radio frequency power as heat, while the antenna radiates it as useful radio waves.

Radiation resistance is very important. Inefficiency in the antenna system appears as a loss resistance in series with the radiation resistance. As antenna length is reduced, as in mobile installations, radiation resistance drops. The radiation resistance of a very short antenna will be less than 5 ohms, comparable to circuit loss resistance. This, and the natural difficulty of feeding it, is why a short antenna cannot be efficient.

12

What is a ground plane?

A ground plane is a large metal-reflecting surface built near, or as part of, an antenna to improve its performance. The reflecting screen under a Marconi antenna is a ground plane. An old application of the ground plane idea is the "counter-poise," a wire fan or screen placed near or on the earth (Fig. 2-7). It serves as an area to germinate the lines of force of the antenna and to carry the resulting currents with less loss than if the earth were used as a ground. This may be unavoidably necessary in a very dry earth or desert installation. From its center to its edge, a ground plane should extend at least a quarter wavelength.

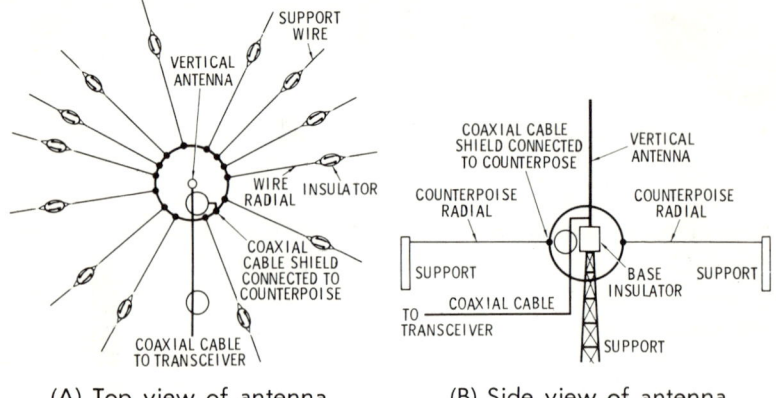

(A) Top view of antenna and counterpoise.

(B) Side view of antenna and counterpoise.

Fig. 2-7. A counterpoise used with a Marconi antenna.

13

What is a loaded antenna?

If an antenna is trimmed or stretched to other than its natural resonant dimensions, it becomes reactive. It acts as if an unwanted capacitor or inductor is added to its circuit. An antenna is "loaded" by adding inductive reactance to balance capacitive reactance or capacitive reactance to balance inductive reactance. This returns the antenna input character to resistive again, even though its length is no longer naturally resonant at the operating frequency.

The most frequently seen loaded antenna is one that is cut shorter than a quarter wave (Fig. 2-8). Usually this antenna is used for a mobile application. Such a shortened Hertz antenna shows unwanted capacitive reactance. Adding a series inductor cancels this so that the transmitter and feed line "see" only the radiation resistance of the antenna.

Fig. 2-8. Base-loaded antenna.

Courtesy Mosley Electronics

14

What is antenna gain?

An apparent transmitter power increase can be achieved by changing antenna design. Usually, antenna gain is rated by comparing its performance to the performance of a dipole under the same conditions. If replacing a dipole by a gain antenna has the same effect as doubling transmitter power, antenna gain is said to be 3 dB.

It makes little difference if the gain antenna is at the transmitting or receiving end of the radio link. If it is at the transmitting end, it boosts effective radiated power by ten times; this is like making the transmitter ten times more powerful. But the same antenna, moved to the receiving end, now picks up ten times more transmitter power, which also is like receiving from a ten times more powerful transmitter.

The properties of a gain antenna can be very good on the CB frequencies. A gain antenna reduces general interference by radiating power in a useful direction rather than upward and/or all around generally. And its response to signals from a preferred direction, when it is directional, reduces interference from other directions.

15

How is antenna gain achieved?

Transmitting antenna gain is achieved by radiating power selectively in a preferred direction, with reduced power going in other directions. Receiving antenna gain is achieved by designing the antenna to pick up more of the incoming radio signal. The same design principles work in both cases. The electric and magnetic fields simply move in opposite directions, rather like a strip of movie film that runs forward or backward through the projector.

An example of a simple gain antenna is a dipole placed at a critical distance in front of a reflecting screen. Part of the emitted radiation travels forward into space. Another part travels back against the screen, is reflected, and returns to reinforce the forward wave. Forward radiation is strengthened at the expense of radiation to the back.

16

What is antenna beam width?

A gain antenna concentrates its transmitted power in certain directions, as shown in Fig. 2-9. The curved lines could have been made by a man walking around the antenna with a field-strength meter, making a map of all the points where he finds the same signal strength. The chart shows some directions where he came very near the antenna, and other points where he had to go very far from it.

The transmitted signal is strongest where he walks within the "lobe," weakest within the notch in the pattern, which is called a "null." The largest or major lobe is the most effective transmitting direction (and receiving direction), and its width is called the antenna beam width.

Fig. 2-9. Example of horizontal radiation patterns.

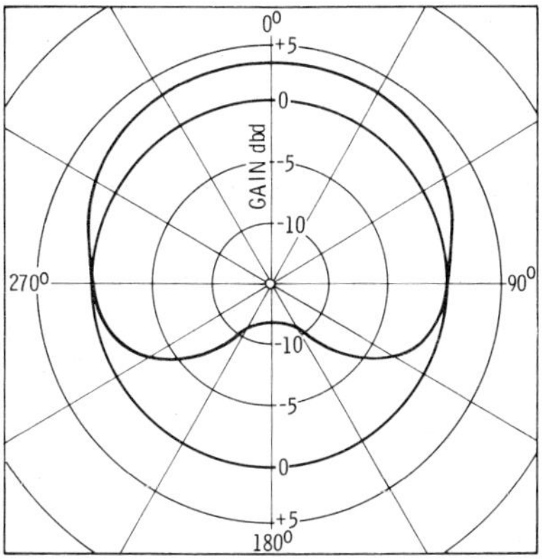

17

What is antenna capture area?

The electric and magnetic fields near an antenna are closely coupled to the antenna. From a radio viewpoint, it is not clear where the antenna ends and free space begins. A receiving antenna occupies far more electrical space than physical space. The electrical space it occupies, as seen by an incoming signal, is called its "capture area."

Capture area is usually estimated in square meters (a meter is about 1.1 yards). Since it is easy to estimate the strength of a direct-wave signal from a distant transmitter when the antenna capture area is known, this incoming signal strength can be converted to an estimate of the signal voltage fed to the antenna terminals of the receivers.

The equation for an incoming power density estimate is:

$$P = 1.64 P_t / 4R^2$$

where,
P is watts per square meter at the receiving antenna,
P_t is watts fed into a dipole at the transmitter,
R is the range in meters.

For instance, it's easy to estimate that at 1 mile (a mile is 1600 meters) from a dipole fed 1 watt of radio power, the power density will be 5 microwatts per square meter.

The equation for the capture area of a receiving dipole is:

$$A = 1.64^2/\lambda$$

where,
A is capture area in square meters,
λ is wavelength in meters.

On CB frequencies, this works out to 16 square meters, which is roughly half a half wave squared. The distant transmitter signal couples 5 microwatts per square meter times 16 square meters or 80 microwatts into the dipole. Since the dipole uses 72-ohm line, and remembering $W = E^2/R$, 80 microwatts appear at the receiver antenna terminals as 76,000 microvolts—a very large signal.

This estimation is dependent on many factors. The transmitting or receiving dipoles might not be correctly oriented or adjusted; transmission lines could be lossy; transmitter or receiver matching could be off; the signal path could be inefficient; or worse yet, there could be two signal paths whose signals arrived out of phase. Yet this result indicates that there's a large margin for signal degredation available here. A watt is far more than adequate for one-mile communication.

Part 3

Antenna and Site Performance

Whenever a traveling radio field intercepts some electrically conducting material, some of the field power is coupled into the conductor. This power is partly lost as heat, and partly reradiated into space. If the conductor is a receiving antenna, much of the intercepted power is fed to the transmission line and receiver. Where the field intercepts the earth, metal building framing, or much electrical wiring, the power is lost.

When the conductor is near a transmitting or receiving antenna, the efficiency of the antenna may be seriously affected by a strong mutual interaction. An engineer's rule of thumb is that anything less than 10 wavelengths from an antenna is "near" it. This works out to about 360 feet at CB frequencies.

Once you understand something about antenna and site performance, you can plan to minimize the unwanted effects. Maybe you can even make the site work for you.

18

How do radio waves travel from one place to another?

Radio waves travel from one place to another by sky wave, direct wave, reflected wave, and ground wave. Signals may travel by two or more routes from a transmitting antenna to the same receiving antenna. Severe fading results from this "multipath" transmission where the two or more different signals arrive at comparable strengths. At CB frequencies, the most common multipath transmission is by direct and reflected waves.

Direct Wave—This is "line-of-sight" transmission, which is more effective and reliable than any other transmission if it can be used. The 10-watt transmitter of the Mariner 4 was heard from the vicinity of Mars and again from the far side of the sun by direct wave. Most CB communication is by direct wave.

Reflected Wave—This is the same as direct wave, except that there are one or more reflecting surfaces along the way. A flashlight beam reflected around a corner is a reflected wave. The reflector may be installed deliberately to get a reflected

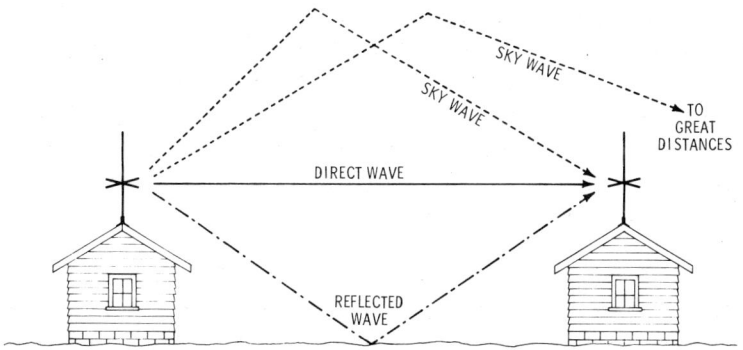

Fig. 3-1. Sky-wave, direct-wave, and reflected-wave propagation.

wave signal into an area where it would otherwise be uselessly weak or absent.

Sky Wave—A signal radiated from a transmitting antenna to the earth's ionosphere, and reflected back down again, as shown in Fig. 3-1 is a sky wave. At CB frequencies, sky wave transmission is only accomplished by signals radiated less than 18 degrees above the true horizon of the earth. The signals arrive at the receiving antenna at similar low angles. Since these angles are in the same ballpark as for direct-wave local work, the antenna cannot select against sky wave signals.

Ground Wave—This is radiation along the earth, which is actually coupled into the earth. At very low frequencies—10 to 30 kHz—powerful transmitters communicate reliably over thousands of miles by ground wave. Since the earth is very lossy at CB frequencies, ground wave transmission is practically useless.

19

What is the ionosphere?

Far beyond our atmosphere, space has perfect radio properties. Near the earth our atmosphere is nearly as good, at CB frequencies. But at 60 to 300 miles up, the air is very thin and it is ionized by the ultraviolet radiation of the sun. This ionized region of several distinct layers, known as the Kennelly-Heaviside layers, has very important radio properties.

It is often called the "ionosphere." Low radio frequencies are absorbed by it. Above the a-m broadcast band is a frequency range of efficient radio reflection, the short-wave region. Near CB frequencies the ionosphere loses most of its reflecting properties, and radio signals pass through. At times the effectiveness of the ionosphere improves and CB signals are reflected at very flat angles as "skip transmission."

It's interesting to know that a primitive radar set was working as early as 1929. The Bureau of Standards used it to measure ionosphere height by transmitting a short radio pulse and timing the return echo. This system was called an "ionospheric sounder."

20

What is site noise?

Site noise is radio noise originating near the receiver site, which is audible when there is no incoming signal. It is also audible when the incoming signal is very weak, and may limit range or reliability at that site. But the site isn't really necessary. In mobile installations, ignition and tire static interference are site problems. Electrical machinery and fluorescent lamps sometimes generate severe site interference, and so do some TV sets.

Since machines not designed to produce radio signals generally radiate little energy, a relatively small move may solve a tough site-noise problem. The move may not work so well if the move merely relocates the radio receiver along a power line carrying objectionable interference.

A special site noise results from corona discharge of natural electricity. This discharge, known to sailors as "St. Elmo's fire," is frequently but not always associated with thunderstorms. It may appear in clear weather, and also snow or dust tend to be very electrical. The resulting interference can be severe and persistent.

At least one CB antenna is supplied with a pair of small metal hoops at its upper end. This is an effective way to reduce corona discharge and the resulting noise. Such an arrangement might also serve to improve the low-angle radiation of the antenna if the engineer includes other appropriate design details.

21

What is a decibel?

A decibel is an engineering unit for comparing relative signal powers or sound volume levels. The decibel system is used because it relates to the natural response of the ear to signal power changes. Choose a signal of any definite power level or loudness. Then another signal at the same level is at zero dB compared to the first, or reference, signal. That is, zero dB is zero difference from the reference level.

Altering the second signal level enough to produce a just detectable loudness difference changes its strength by 1 decibel. This turns out, upon power measurement, to be a 25 percent power change as an increase or a decrease. A 3 dB change is easier to hear, and corresponds to a doubling or a halving of power. Finally, a distinct power level change which the ear rates as twice or half as loud as the reference level is about a 10 dB change. This turns out to be a power change of ten times. And 20 dB is a one-hundred times power change.

The decibel system is also very useful for engineering. It provides a way to compare large power level differences without using large numbers. For instance, a million-times power gain is 60 dB. The overall power gain of a good receiver when it is receiving a weak signal will exceed 120 dB, or one million million.

All decibel measurements consist of two parts: reference level, and a specification of decibels difference from the reference level. Often the reference level isn't stated. For instance, the noise level in a room may be 60 dB. Relative to what? It turns out that the audio industry conventionally rates

Table 3-1. Decibel Table

dB	Current or Voltage Ratio		Power Ratio		dB	Current or Voltage Ratio		Power Ratio	
	Gain	Loss	Gain	Loss		Gain	Loss	Gain	Loss
0	1.000	1.0000	1.000	1.0000	5.0	1.778	.5623	3.162	.3162
.1	1.012	.9886	1.023	.9772	5.1	1.799	.5559	3.236	.3090
.2	1.023	.9772	1.047	.9550	5.2	1.820	.5495	3.311	.3020
.3	1.035	.9661	1.072	.9333	5.3	1.841	.5433	3.388	.2951
.4	1.047	.9550	1.096	.9120	5.4	1.862	.5370	3.467	.2884
.5	1.059	.9441	1.122	.8913	5.5	1.884	.5309	3.548	.2818
.6	1.072	.9333	1.148	.8710	5.6	1.905	.5248	3.631	.2754
.7	1.084	.9226	1.175	.8511	5.7	1.928	.5188	3.715	.2692
.8	1.096	.9120	1.202	.8318	5.8	1.950	.5129	3.802	.2630
.9	1.109	.9016	1.230	.8128	5.9	1.972	.5070	3.890	.2570
1.0	1.122	.8913	1.259	.7943	6.0	1.995	.5012	3.981	.2512
1.1	1.135	.8810	1.288	.7762	6.1	2.018	.4955	4.074	.2455
1.2	1.148	.8710	1.318	.7586	6.2	2.042	.4898	4.169	.2399
1.3	1.161	.8610	1.349	.7413	6.3	2.065	.4842	4.266	.2344
1.4	1.175	.8511	1.380	.7244	6.4	2.089	.4786	4.365	.2291
1.5	1.189	.8414	1.413	.7079	6.5	2.113	.4732	4.467	.2239
1.6	1.202	.8318	1.445	.6918	6.6	2.138	.4677	4.571	.2188
1.7	1.216	.8222	1.479	.6761	6.7	2.163	.4624	4.677	.2138
1.8	1.230	.8128	1.514	.6607	6.8	2.188	.4571	4.786	.2089
1.9	1.245	.8035	1.549	.6457	6.9	2.213	.4519	4.898	.2042
2.0	1.259	.7943	1.585	.6310	7.0	2.239	.4467	5.012	.1995
2.1	1.274	.7852	1.622	.6166	7.1	2.265	.4416	5.129	.1950
2.2	1.288	.7762	1.660	.6026	7.2	2.291	.4365	5.248	.1905
2.3	1.303	.7674	1.698	.5888	7.3	2.317	.4315	5.370	.1862
2.4	1.318	.7586	1.738	.5754	7.4	2.344	.4266	5.495	.1820
2.5	1.334	.7499	1.778	.5623	7.5	2.371	.4217	5.623	.1778
2.6	1.349	.7413	1.820	.5495	7.6	2.399	.4169	5.754	.1738
2.7	1.365	.7328	1.862	.5370	7.7	2.427	.4121	5.888	.1698
2.8	1.380	.7244	1.905	.5248	7.8	2.455	.4074	6.026	.1660
2.9	1.396	.7161	1.950	.5129	7.9	2.483	.4027	6.166	.1622
3.0	1.413	.7079	1.995	.5012	8.0	2.512	.3981	6.310	.1585
3.1	1.429	.6998	2.042	.4898	8.1	2.541	.3936	6.457	.1549
3.2	1.445	.6918	2.089	.4786	8.2	2.570	.3890	6.607	.1514
3.3	1.462	.6839	2.138	.4677	8.3	2.600	.3846	6.761	.1479
3.4	1.479	.6761	2.188	.4571	8.4	2.630	.3802	6.918	.1445
3.5	1.496	.6683	2.239	.4467	8.5	2.661	.3758	7.079	.1413
3.6	1.514	.6607	2.291	.4365	8.6	2.692	.3715	7.244	.1380
3.7	1.531	.6531	2.344	.4266	8.7	2.723	.3673	7.413	.1349
3.8	1.549	.6457	2.399	.4169	8.8	2.754	.3631	7.586	.1318
3.9	1.567	.6383	2.455	.4074	8.9	2.786	.3589	7.762	.1288
4.0	1.585	.6310	2.512	.3981	9.0	2.818	.3548	7.943	.1259
4.1	1.603	.6237	2.570	.3890	9.1	2.851	.3508	8.128	.1230
4.2	1.622	.6166	2.630	.3802	9.2	2.884	.3467	8.318	.1202
4.3	1.641	.6095	2.692	.3715	9.3	2.917	.3428	8.511	.1175
4.4	1.660	.6026	2.754	.3631	9.4	2.951	.3388	8.710	.1148
4.5	1.679	.5957	2.818	.3548	9.5	2.985	.3350	8.913	.1122
4.6	1.698	.5888	2.884	.3467	9.6	3.020	.3311	9.120	.1096
4.7	1.718	.5821	2.951	.3388	9.7	3.055	.3273	9.333	.1072
4.8	1.738	.5754	3.020	.3311	9.8	3.090	.3236	9.550	.1047
4.9	1.758	.5689	3.090	.3236	9.9	3.126	.3199	9.772	.1023

Table 3-1. Decibel Table—Cont

dB	Current or Voltage Ratio		Power Ratio		dB	Current or Voltage Ratio		Power Ratio	
	Gain	Loss	Gain	Loss		Gain	Loss	Gain	Loss
10.0	3.162	.3162	10.000	.1000	15.0	5.623	.1778	31.62	.03162
10.1	3.199	.3126	10.23	.09772	15.1	5.689	.1758	32.36	.03090
10.2	3.236	.3090	10.47	.09550	15.2	5.754	.1738	33.11	.03020
10.3	3.273	.3055	10.72	.09333	15.3	5.821	.1718	33.88	.02951
10.4	3.311	.3020	10.96	.09120	15.4	5.888	.1698	34.67	.02884
10.5	3.350	.2985	11.22	.08913	15.5	5.957	.1679	35.48	.02818
10.6	3.388	.2951	11.48	.08710	15.6	6.026	.1660	36.31	.02754
10.7	3.428	.2917	11.75	.08511	15.7	6.095	.1641	37.15	.02692
10.8	3.467	.2884	12.02	.08318	15.8	6.166	.1622	38.02	.02630
10.9	3.508	.2851	12.30	.08128	15.9	6.237	.1603	38.90	.02570
11.0	3.548	.2818	12.59	.07943	16.0	6.310	.1585	39.81	.02512
11.1	3.589	.2786	12.88	.07762	16.1	6.383	.1567	40.74	.02455
11.2	3.631	.2754	13.18	.07586	16.2	6.457	.1549	41.69	.02399
11.3	3.673	.2723	13.49	.07413	16.3	6.531	.1531	42.66	.02344
11.4	3.715	.2692	13.80	.07244	16.4	6.607	.1514	43.65	.02291
11.5	3.758	.2661	14.13	.07079	16.5	6.683	.1496	44.67	.02239
11.6	3.802	.2630	14.45	.06918	16.6	6.761	.1479	45.71	.02188
11.7	3.846	.2600	14.79	.06761	16.7	6.839	.1462	46.77	.02138
11.8	3.890	.2570	15.14	.06607	16.8	6.918	.1445	47.86	.02089
11.9	3.936	.2541	15.49	.06457	16.9	6.998	.1429	48.98	.02042
12.0	3.981	.2512	15.85	.06310	17.0	7.079	.1413	50.12	.01995
12.1	4.027	.2483	16.22	.06166	17.1	7.161	.1396	51.29	.01950
12.2	4.074	.2455	16.60	.06026	17.2	7.244	.1380	52.48	.01905
12.3	4.121	.2427	16.98	.05888	17.3	7.328	.1365	53.70	.01862
12.4	4.169	.2399	17.38	.05754	17.4	7.413	.1349	54.95	.01820
12.5	4.217	.2371	17.78	.05623	17.5	7.499	.1334	56.23	.01778
12.6	4.266	.2344	18.20	.05495	17.6	7.586	.1318	57.54	.01738
12.7	4.315	.2317	18.62	.05370	17.7	7.674	.1303	58.88	.01698
12.8	4.365	.2291	19.05	.05248	17.8	7.762	.1288	60.26	.01660
12.9	4.416	.2265	19.50	.05129	17.9	7.852	.1274	61.66	.01622
13.0	4.467	.2239	19.95	.05012	18.0	7.943	.1259	63.10	.01585
13.1	4.519	.2213	20.42	.04898	18.1	8.035	.1245	64.57	.01549
13.2	4.571	.2188	20.89	.04786	18.2	8.128	.1230	66.07	.01514
13.3	4.624	.2163	21.38	.04677	18.3	8.222	.1216	67.61	.01479
13.4	4.677	.2138	21.88	.04571	18.4	8.318	.1202	69.18	.01445
13.5	4.732	.2113	22.39	.04467	18.5	8.414	.1189	70.79	.01413
13.6	4.786	.2089	22.91	.04365	18.6	8.511	.1175	72.44	.01380
13.7	4.842	.2065	23.44	.04266	18.7	8.610	.1161	74.13	.01349
13.8	4.898	.2042	23.99	.04169	18.8	8.710	.1148	75.86	.01318
13.9	4.955	.2018	24.55	.04074	18.9	8.811	.1135	77.62	.01288
14.0	5.012	.1995	25.12	.03981	19.0	8.913	.1122	79.43	.01259
14.1	5.070	.1972	25.70	.03890	19.1	9.016	.1109	81.28	.01230
14.2	5.129	.1950	26.30	.03802	19.2	9.120	.1096	83.18	.01202
14.3	5.188	.1928	26.92	.03715	19.3	9.226	.1084	85.11	.01175
14.4	5.248	.1905	27.54	.03631	19.4	9.333	.1072	87.10	.01148
14.5	5.309	.1884	28.18	.03548	19.5	9.441	.1059	89.13	.01122
14.6	5.370	.1862	28.84	.03467	19.6	9.550	.1047	91.20	.01096
14.7	5.433	.1841	29.51	.03388	19.7	9.661	.1035	93.33	.01072
14.8	5.495	.1820	30.20	.03311	19.8	9.772	.1023	95.50	.01047
14.9	5.559	.1799	30.90	.03236	19.9	9.886	.1012	97.72	.01023

Note: For values from 20 to 180 db, see next page.

Table 3-1. Decibel Table—Cont

dB	Current or Voltage Ratio		Power Ratio	
	Gain	Loss	Gain	Loss
20.0	10.00	0.1000	100.00	0.01000
25.0	17.78	0.0562	3.162×10^2	3.162×10^{-3}
30.0	31.62	0.0316	10^3	10^{-3}
35.0	56.23	0.0178	3.162×10^3	3.162×10^{-4}
40.0	100.00	0.0100	10^4	10^{-4}
45.0	177.8	0.0056	3.162×10^4	3.162×10^{-5}
50.0	316.2	0.0032	10^5	10^{-5}
55.0	562.3	0.0018	3.162×10^5	3.162×10^{-6}
60.0	10^3	10^{-3}	10^6	10^{-6}
65.0	1.778×10^3	5.623×10^{-4}	3.162×10^6	3.162×10^{-7}
70.0	3.162×10^3	3.162×10^{-4}	10^7	10^{-7}
75.0	5.623×10^3	1.78×10^{-4}	3.162×10^7	3.162×10^{-8}
80.0	10^4	10^{-4}	10^8	10^{-8}
85.0	1.778×10^4	5.623×10^{-5}	3.162×10^8	3.162×10^{-9}
90.0	3.162×10^4	3.162×10^{-5}	10^9	10^{-9}
95.0	5.632×10^4	1.78×10^{-5}	3.162×10^9	3.162×10^{-10}
100.0	10^5	10^{-5}	10^{10}	10^{-10}
110.0	3.162×10^5	3.162×10^{-6}	10^{11}	10^{-11}
120.0	10^6	10^{-6}	10^{12}	10^{-12}
130.0	3.162×10^6	3.162×10^{-7}	10^{13}	10^{-13}
140.0	10^7	10^{-7}	10^{14}	10^{-14}
150.0	3.162×10^7	3.162×10^{-8}	10^{15}	10^{-15}
160.0	10^8	10^{-8}	10^{16}	10^{-16}
170.0	3.162×10^8	3.162×10^{-9}	10^{17}	10^{-17}
180.0	10^9	10^{-9}	10^{18}	10^{-18}

sound levels against the weakest sound an average ear can detect. A room sound level of 60 dB is very quiet—and a million times louder than the best sensitivity of the ear.

Another decibel measurement describes receiver performance. For instance, at a certain antenna input signal level the receiver output is reported to be 10 dB S/N ratio. This means that the audio output is ten times as powerful as the receiver internal noise for this level of input signal.

Its "sliding scale" effect makes decibel measurement a little harder to follow than voltage or current testing (see Table 3-1). But it is very well adapted to real life communications measurement problems.

What is an "S" unit?

The "S" unit is based on an old commercial and radio amateur system of reporting received signal strength. A signal that is barely detectable is said to be "S1." When it is extremely strong and clear, it is reported to be "S9." People even report such readings as "20 dB over S9," but from an engineering viewpoint this has no real meaning.

Table 3-2. "S" Meter Indications

1	Barely perceptible		6	Good
2	Very weak		7	Quite good
3	Weak		8	Loud and clear
4	Fair		9	Excellent
5	Fairly good			

This is because the "S" unit isn't defined in the way a volt, ampere, or watt is defined. The "S" unit system simply reports what the listener hears. If he readjusts his receiver, finds a better receiver, or improves his antenna, he can then give a much better report on the same strength signal coming into the same operating position. There is no agreement among manufacturers or engineers what an "S" unit represents in terms of signal power. Various manufacturers and magazines publish "S" unit charts from time to time. Probably the best way to use the "S" unit system is the way it was originally used, as listed in Table 3-2.

23

How does effective antenna elevation affect range?

Communication range and reliability depend upon several factors. If elevation is one of them, then at a given site a higher antenna will outperform a lower one. And an antenna at a higher site may outperform another at a nearby lower site. But sometimes the lower antenna will perform comparably, or even better (Fig. 3-2).

Where there are no hills, buildings, and other interfering objects, direct-wave signals are usable to the radio horizon. If the topography is such that raising the antenna does not extend the radio horizon, then raising the antenna will not affect maximum range. But it may improve effectiveness at shorter ranges in hilly or rolling country. Where conditions are such that raising the transmitting antenna improves communication, it may help further to raise the receiving antenna also.

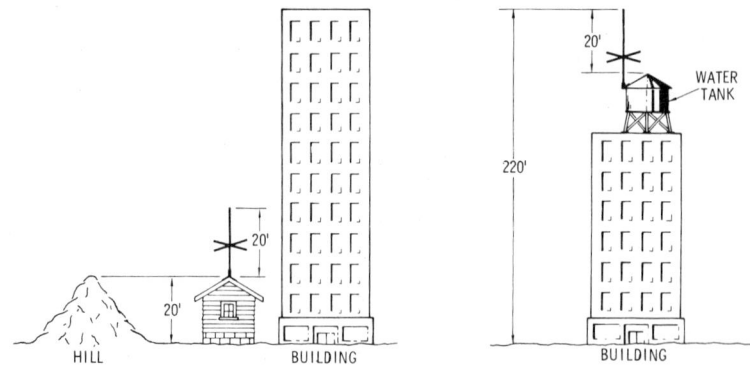

(A) Effective antenna elevation is only 20 feet, and only to the left, even if antenna is 40 feet above ground.

(B) Effective antenna elevation is 220 feet although antenna is at 20-foot legal height.

Fig. 3-2. Examples of effective antenna elevation.

Under certain conditions, a lower receiving antenna may out-perform a higher one. If direct and reflected waves arrive with similar strengths but out of phase at the receiving antenna, severe fading results. Lowering the antenna will reduce the phase difference between the two incoming signals, and may produce a remarkable signal strength improvement. Relocating the antenna only a few feet in any direction can also be very effective in this case.

24

When should vertical polarization be used?

If one station uses vertical polarization, another communicating with it often finds that vertical polarization works better than horizontal polarization for that contact. An example is a mobile car or boat installation, where a vertically polarized antenna is nearly always used simply for mechanical convenience or necessity. Vertically polarized antennas are said to be slightly more susceptible than horizontally polarized antennas to ignition and other man-made interference. The advantages of vertical over horizontal polarizations or horizontal over vertical polarizations are small enough to have produced a lively controversy over several years among radio amateurs.

One reason for choosing vertical polarization has some engineering strength. A horizontal dipole type antenna is bidirectional and radiates much power at high angles, if it is not placed a multiple of a half wave above good earth or a large ground plane. All things considered, this is hard to do. But the vertical radiation of a vertical dipole is near zero, and the horizon gets the strongest signal. Where vertical polarization has any clear advantage at CB frequencies, this is it. Its actual radiation of vertically polarized radio waves is a by-product of the most effective antenna design.

Also, since most mobile CB antennas are vertically polarized and omnidirectional, reception is essentially equal from all directions.

25

When should horizontal polarization be used?

Horizontal polarization works best when communicating with another station using a horizontally polarized antenna. Site characteristics are one reason for choosing such an antenna design.

Actually, antenna polarization is not very meaningful in many near-the-earth applications. If a radio signal is reflected by a surface that is not square to its polarization, the reflection changes its polarization. And many antennas that appear horizontally polarized, such as the dipole, only radiate a horizontally polarized signal in certain directions. The strong advocate of one polarization over another should make certain the polarization he is advocating is the same as he is using.

Polarizations available by choosing different antenna designs are tilted as well as horizontal or vertical, and right-hand or left-hand elliptical and circular polarizations.

26

What is a passive repeater?

A passive repeater is a very simple installation designed to reflect or project a radio signal into an area where it would otherwise be inaudible. A passive repeater has no electronic parts, uses no power, and doesn't need FCC licensing. There are two general types.

The simplest is a large metal sheet or screen that acts like a mirror. It reflects incoming radio waves up to 90 degrees from their incoming course. Beyond 90 degrees, reflector size increases rapidly because little reflection is seen by the incoming signal, and, therefore, little signal is reflected. The flat metal screen on microwave towers are passive repeaters reflecting the radiation from parabolic antennas near the ground.

The reflector idea can be improved upon by using antennas. One antenna, oriented for maximun power pickup, is aimed at or visible to the transmitter, as shown in Fig. 3-3. The other is aimed at the receiver site, and it may be some distance from

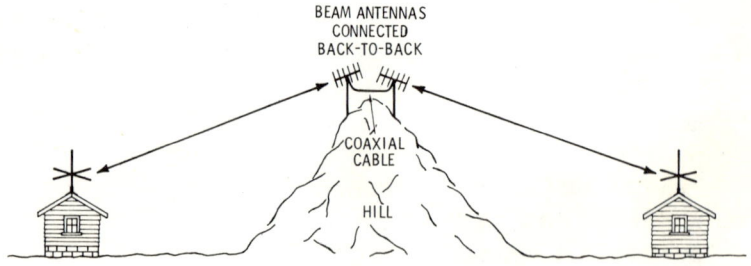

Fig. 3-3. Passive repeater.

the first. The antennas are interconnected with good quality transmission line. This system can get into deep valleys or around mountains. Like the simpler reflector, it works equally well both ways.

27

Is it lawful to use a gain antenna?

Yes, if the gain antenna gets its gain by reducing radiation in some directions to emphasize it in others. Gain antennas of this type are not only lawful, but often desirable. They reduce the contribution of a station to band congestion by concentrating its output in a useful direction. And, the sensitivity of the station to outside interference is also reduced.

But if the gain antenna gets its gain from a built-in transmitting power amplifier that exceeds the FCC's 5-watt limit, then the gain antenna is not lawful. Whether situated at the operating position or placed in the antenna structure, a power amplifier is illegal if overpowered or unlicensed.

28

How can two or more antennas be used?

From an engineering viewpoint, the use of a single antenna at an installation may be very restrictive. The directivity of a gain antenna may be unwanted in some conditions, while the lack of directional control of an omnidirectional antenna works poorly in others (Fig. 3-4). One of each is a good mix, and further variation is possible if one antenna is restricted to receiving only. This antenna can be located without regard to the Part 95 regulations, except for the 1 per 200-foot airport rule.

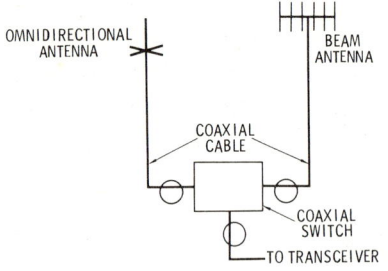

Fig. 3-4. More than one antenna may be used at a CB station.

It is a basic radio engineering fact that if the antennas at both ends of a radio link cannot be placed very high, but the situation is one where height is desirable, it helps considerably to place one antenna high. A 20-foot transmitting antenna and a 100-foot receiving antenna might be rather extreme, but it is good engineering—and it is legal (Fig. 3-5).

Another multiple-antenna arrangement simply has two or more antennas fixed in place and adjusted for most effective communication with other definite locations. This makes good sense, too, because each antenna can be placed where it gives the best results over a particular fixed communications path. Multiple antenna installations are limited only by the installer's ingenuity and capital.

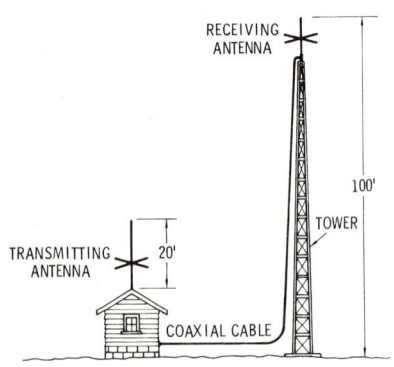

Fig. 3-5. The 20-foot height restriction does not apply to the receiving antenna.

29

Can the same antenna be used for two or more transceivers?

Yes, for simultaneous reception, but not for simultaneous transmission. The transceiver inputs cannot simply be paralleled to the antenna system. Special couplers are required. The station is licensed, not the antenna. If the transceivers are legal, the antenna is legal for any one of them.

30

What is an antenna amplifier?

An antenna amplifier is a small-signal receiving-only amplifier built into the antenna structure. This is an excellent idea from an engineering viewpoint, since transmission line losses are eliminated and other losses are minimized. Commercial antenna amplifier designs provide for removing the amplifier from the antenna circuit when transmitting.

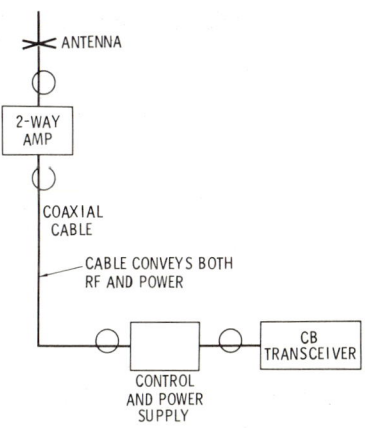

Fig. 3-6. Two-way antenna amplifier improves receiving sensitivity and makes up for transmission coaxial cable losses.

Where good components and low-noise transistors are used, such an amplifier improves sensitivity to a point where receiving performance is limited only by irreducible outside noise. An antenna amplifier is not in order for congested urban and surburban areas. In these regions, station effectiveness is determined by selectivity and resistance to intermodulation, not by sheer sensitivity.

A two-way CB amplifier has been developed which amplifies both received signals and transmitted signals. Power fed to the antenna is boosted to, and limited to, 4 watts. It is connected as shown in Fig. 3-6.

31

Can a CB transceiver be mounted near the antenna but controlled from a distance?

Yes. If the antenna is very near the operating position, there is no possibility that an unauthorized party might gain control of the transmitter, and the transmitter performance is

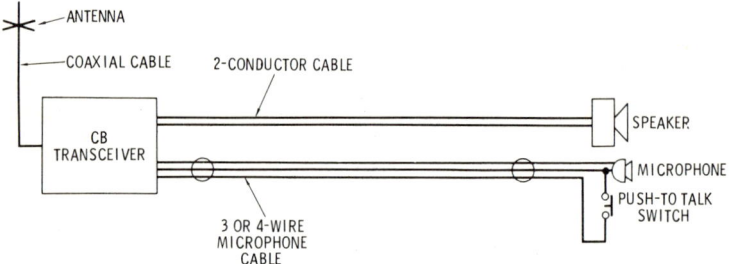

Fig. 3-7. An example of extended local control.

legal and will stay legal. The Part 95 regulations do not state that the transmitter must be at the operating position. However, both must be on the same premises.

The transceiver can be connected to a remote microphone, loudspeaker, and transmitter control through multiconductor cable, as shown in Fig. 3-7.

32

How can a range field survey be conducted?

The first step is to estimate the radio horizon as seen from the transmitter site. A road map is usable, but a topographic map is far better. Landmarks visibly near line-of-sight limits will assist in this. Their locations are plotted on the map, and a rough border is drawn around the transmitter site about 10 percent beyond the estimated visual horizon. This border closes up to the visual limit at the tops of steep hills, and suspected poor-reception areas inside the border can be roughed in too. This gives an overall picture of the usable transmission area without using any testing at all.

The next step is to test these estimates in areas where the borders need to be known accurately. A man with a mobile rig or a walkie-talkie goes to the critical areas to determine the working limits. If his antenna is directional, he always turns its worst side toward the transmitter, in order to get a worst-case distance limit. Since actual results will correspond somewhat to the estimated limit, the mobile station does not need to test all possible directions. Three to six sites per compass quadrant should be adequate. Suspected shadow areas are checked the same way.

WARNING: Test transmissions should be limited to the shortest time and in accordance with Part 95 regulations.

33

Is an earth ground required?

An earth ground is not required when using a conventional CB antenna. An earth ground is needed at lower frequencies when using a Marconi antenna. At CB frequencies, an earth ground isn't needed when it is replaced by something equivalent, like a ground plane or a car body.

If there is a lightning hazard, an earth ground or its maritime equivalent should be provided. A lightning flash that breaks through a half mile of air won't be stopped by any insulating scheme a man can install. The possible stroke and static build-ups must be provided with a simple, direct route to where it's going anyway. A base station antenna may be grounded for lightning protection as shown in Fig. 3-8.

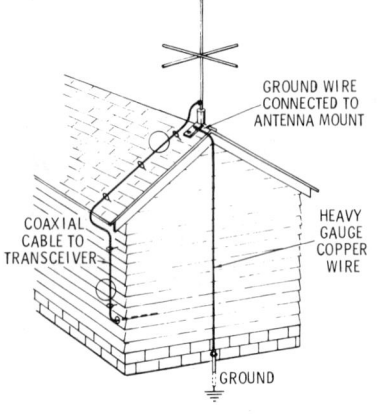

Fig. 3-8. Lightning protection grounding technique.

34

How can skip reception and transmission be prevented?

Skip reception and transmission cannot be prevented. Skip communication is illegal, but can be avoided by not answering calls beyond the 150-mile limit. This problem hasn't an engineering solution because the same antenna geometry that provides the best local communication emphasizes the low-angle radiation optimum for skip. It's entirely up to the operator.

Part 4

Real Antennas

Antenna sizes vary from little printed-circuit assemblies that will fit a shirt pocket to huge nets strung between mountains or "Mills Cross" installations covering square miles of desert floor. Complexities vary from multiple-transmitter systems on skyscrapers to simple folded dipoles put up with no engineering at all.

There are good reasons for this tremendous variation among antennas. Some are special requirements dictated by operating frequency, narrow or wideband operation, patterns with or without nulls, controlled vertical radiation, engineering practicality, and of course, cost. An important reason is that modern science and engineering technology seems able to evolve a new best kind of antenna to meet every new need. An example is the "slot" antenna for high-speed aircraft.

There isn't any one best antenna, not even in the well-defined CB field. The CBer has a wide range of manufactured antennas at all prices to choose from. And he can design and build more special types for himself. Fig. 4-1 illustrates examples of CB base station antennas, and Fig. 4-2 typical mobile antennas.

35

What is a tradeoff?

A tradeoff is a wise engineering choice sacrificing something to get more of something else. The good designer tries to achieve antenna gain, bandwidth, reliability, stability, installation convenience, manufacturing practicality, and other qualities. He tries to manage these in a sensible proportion, balancing one against the next. He makes not one but many tradeoff decisions in designing his antenna. So does the man who simply seeks to install an existing antenna.

For instance, the good installation worker tries to place the antenna where it will produce the best results. He may have to try some tests. Somewhere along the way, he decides more testing isn't worth the possible return. A higher antenna costs more, is harder to install, and is more susceptible to wind

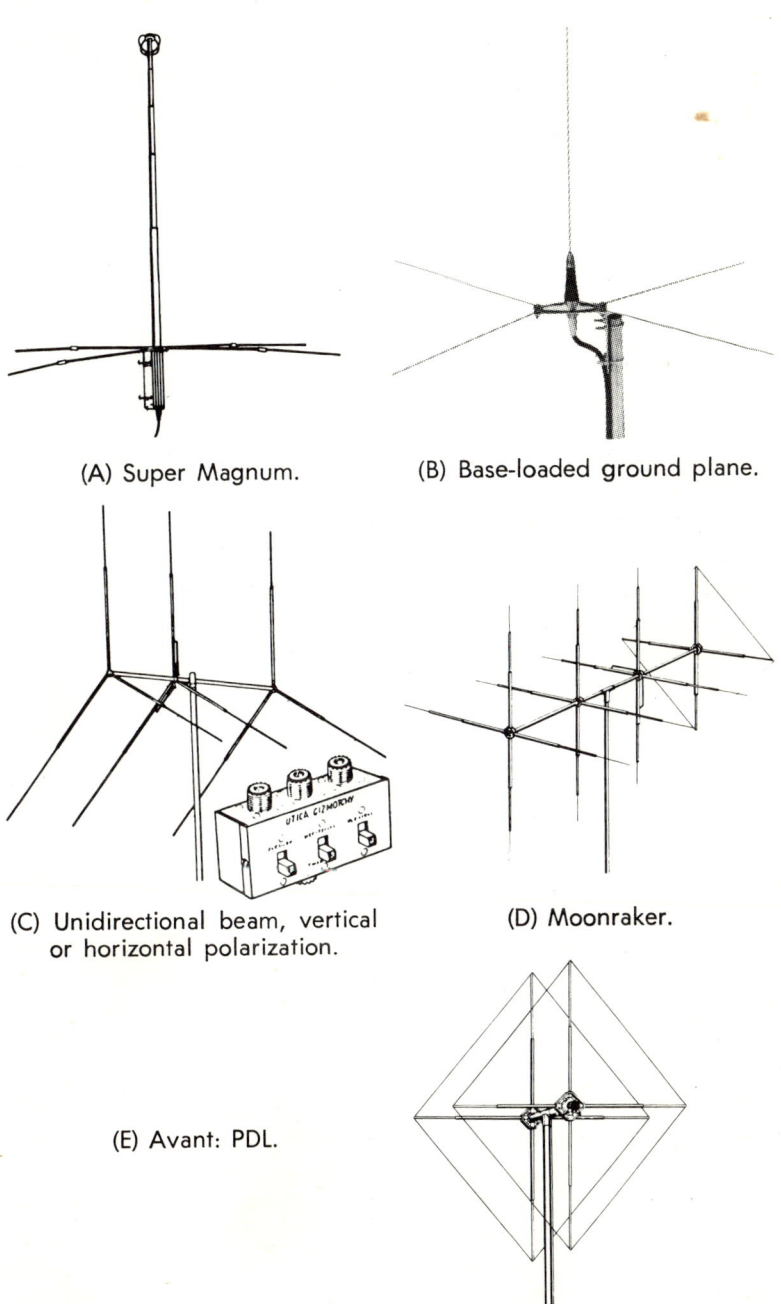

(A) Super Magnum. (B) Base-loaded ground plane.

(C) Unidirectional beam, vertical or horizontal polarization. (D) Moonraker.

(E) Avant: PDL.

Fig. 4-1. Examples of base-station antennas.

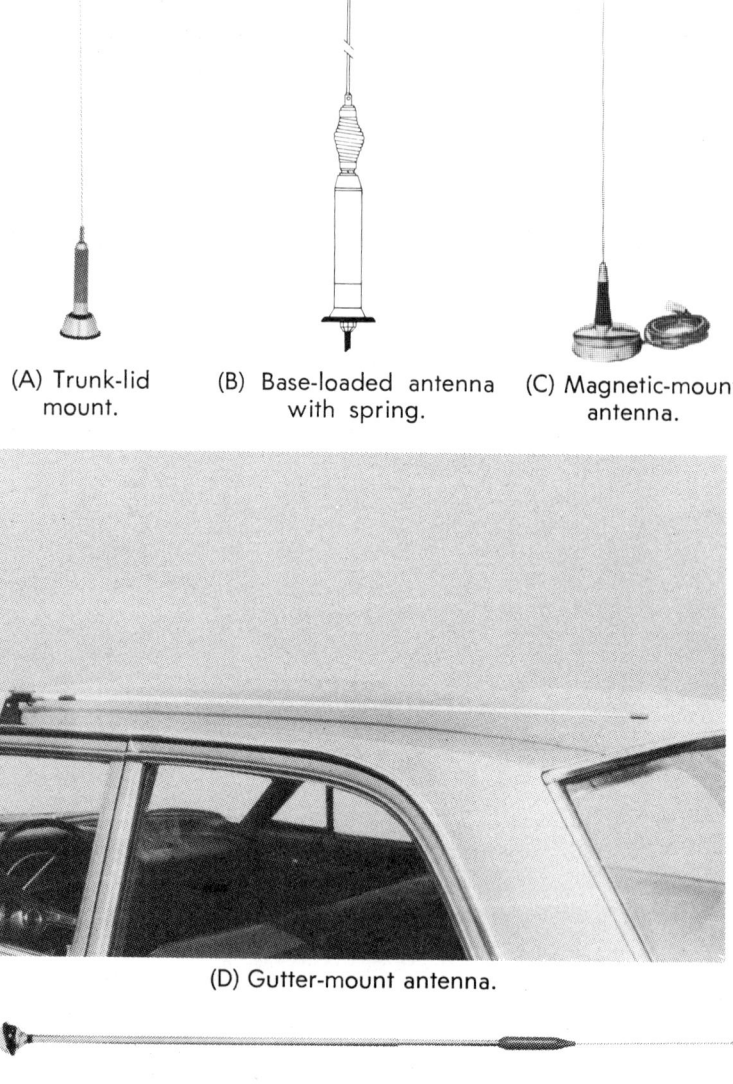

(A) Trunk-lid mount. (B) Base-loaded antenna with spring. (C) Magnetic-mount antenna.

(D) Gutter-mount antenna.

(E) Combination CB/A-M antenna.

(F) Fiberglass whip antenna.
Fig. 4-2. Examples of mobile antennas.

or lightning damage. This could be very important in the southern states where hurricane winds have reached 160 miles per hour and some areas get hundreds of thunderstorms each year. The good technician thinks easily in tradeoff terms.

36

What is antenna bandwidth?

Antenna bandwidth is the frequency range over which the antenna performs satisfactorily without adjustments. Most CB antennas are resonant structures whose performance is sharply frequency dependent. Their bandwidth is only a few percent of the operating frequency. This is satisfactory for CB operation, and is economical since resonant designs offer the most performance for the cost and installation size. Bandwidth is often given in SWR terms as the frequency range over which the feedline SWR remains below 1.5:1.

A simple, unusual antenna, the discone, works well over a 10.1 frequency range. For instance, one discone would serve at an installation working over the 20 to 200 MHz range.

37

What is SWR?

SWR is standing-wave ratio, a measure of the power reflected from a transmission line output end back toward its input. If there is no reflected power, tests along a transmission line will find the same voltage at all points. But when there is power flowing both ways in the line, voltage and current interactions will result in fixed, detectable voltage differences called "loops" and "nodes." The standing-wave ratio of a line is simply the ratio between the largest voltage, at a loop, and the least voltage, at a node, as illustrated in Fig. 4-3.

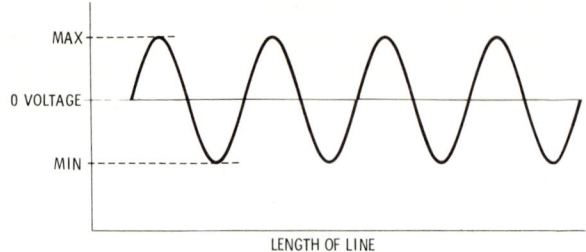

Fig. 4-3. Standing waves on a transmission line.

The common "SWR" meter doesn't actually measure standing-wave ratio. It compares forward and reflected *power*. Its scale is calibrated to indicate the VSWR (voltage standing-wave ratio) that would be found for various percentages of forward power returning from the transmission line load end.

38

What is a Yagi antenna?

Antennas have pedigrees. Knowing their ancestry and family relations helps understand how they work. Much apparent confusion about short-wave antennas can be cleared up by finding their dipole ancestry. Whole chapters of large engineering books deal with antennas very like dipoles assembled into various arrays. All these antennas relate simply to the original physicist's "di-pole" of Question No. 6. A Yagi antenna (Fig. 4-4) is a dipole variant consisting of a driven dipole, a reflector, and one or more director elements, which radiate maximum energy in one direction.

39

What is a dual polarized antenna?

A dual polarized antenna is an antenna that is remotely adjustable to transmit or receive horizontally or vertically polarized waves. In some installations, the change is made by tilting the antenna from vertical to horizontal or back again. In some modern antenna designs, polarization is changed by electrical switching from the operating position.

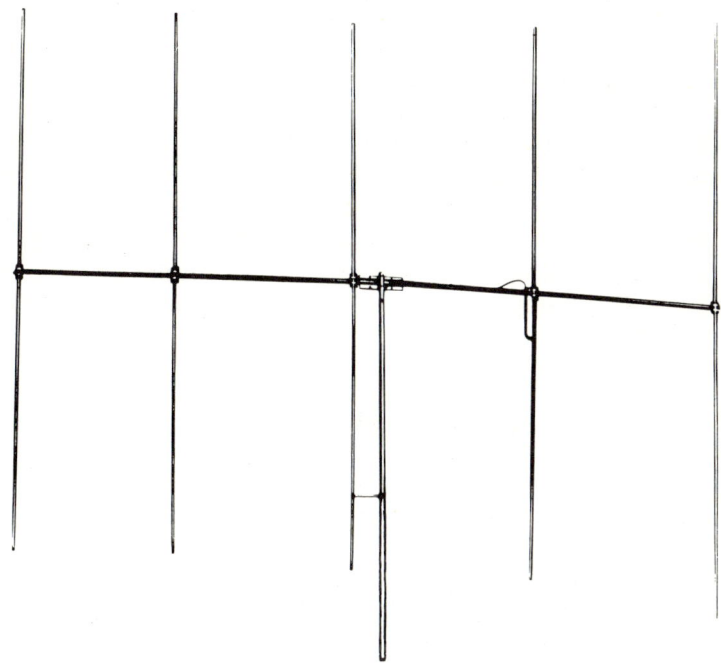

Fig. 4-4. Yagi antenna.

One such arrangement features a vertical radiator and two more radiators at downward angles under it. As a vertically polarized antenna, it works as a modified ground plane resembling a vertical dipole, since it hasn't enough lower elements to make a good ground plane. As a horizontally polarized antenna, its vertical radiator is unused, and the two lower elements act much like a horizontal dipole. Directivity may be improved by adding a second vertical/horizontal assembly, either parasitic or fed, near the first.

40

What is a quad antenna?

A quad antenna is a simple, uncritical, and highly effective design resembling a loop antenna (Fig. 4-5). However, it is actually very different from the loop antenna. It uses space very effectively since its size is a quarter wavelength on a side, or about 4½ feet each side of center at the CB frequencies. Electrically, it is a large wire square fed at the bottom. Working out the voltage distribution around the square reveals that the high voltage points are halfway up the vertical sides of the square. Since the electric lines of force extend across the square, the quad antenna signal is horizontally polarized.

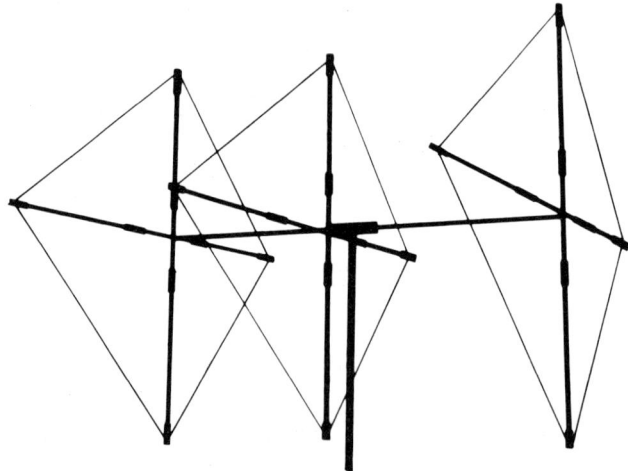

Courtesy Mosley Electronics, Inc.

Fig. 4-5. Cubical quad antenna.

The quad antenna can be made more directive by adding a tuned reflector off one open side of the radiating loop, and even a tuned director loop off the other open side. This is like a Yagi. The additional loops are tuned with small variable capacitors, a great convenience since it isn't necessary to take the assembly down and trim metal parts for tuning.

41

What is a "low-profile" antenna?

A "low-profile" antenna is a special type of antenna designed for use on large vehicles to avoid overhead clearance problems.

42

What is a "disguise" antenna?

A "disguise" antenna is a CB antenna that looks like an ordinary car radio antenna. To further disguise a mobile installation, the CB rig may be installed in the glove compartment or elsewhere out of sight. The likelihood of theft, very great in large cities, is considerably reduced.

43

What is a "scanner-type" antenna?

It is possible to change the directive pattern of an antenna without rotating the antenna. For instance, a two-element radiator/reflector antenna shows maximum gain along a line from the reflector forward through the radiator. If the radiator and reflector terminals are brought down the mast by transmission line, or controlled by relays in the antenna structure, then either antenna element can serve as radiator or reflector. This scheme gives good electrically controlled directivity in two directions.

Using three elements, this scheme becomes much more versatile. One, two, or all three elements can serve as radiators; one or two as reflectors. Many directive patterns are possible.

44

What is a fiberglass antenna?

A fiberglass antenna is a length of fiberglass-reinforced plastic with a piece of wire wound around it or imbedded in it. "Fiberglass antenna" sounds like a contradiction in terms since fiberglass-resin structures have excellent to superior insulating properties. The manufacturer adds a length of wire to the fiberglass structure and winds up with a very satisfactory antenna. It is said to be stronger than steel, and more flexible and chemical resistant. The antenna electrical length is increased by inductive loading if the wire is wound around the shaft in a long spiral (Fig. 4-2F).

45

Can an indoor antenna be used for CB?

An indoor antenna is often used for CB. Its effectiveness will depend upon the site qualities and the nature of the building. Site requirements are the same as for outdoor antenna mounting.

If the building is a wood-framed shed or house, performance will closely resemble that of an antenna in a similar location outdoors nearby. Antenna tuning and performance will be affected by house wiring, and may be badly affected by a metal roof. In a modern steel-frame brick structure, performance will be noticeably reduced. A window-mount antenna would be more satisfactory.

Performance in a reinforced-concrete building is also likely to be extremely poor. Concrete is generally reinforced by several steel rods assembled into a hard, tight mesh before the concrete is poured. This reinforcing makes an effective radio signal screen. If it is necessary to operate in such a site, various antenna polarizations should be tried. Signal loss will be least when the polarization is at right angles to the reinforcing bars.

Performance in a metal-lined room or steel box—like a bank safe—is zero. No practical reception or transmission can be expected in such a site.

46

Can a long-wire antenna be used for CB?

Yes, provided it does not violate the 20-foot antenna height limit imposed by the Part 95 rules. The problem is to couple the transmitter into it. Radio amateur books and magazines contain many schemes for feeding power into long-wire antennas and for determining that the power is being radiated rather than wasted in ohmic resistance. Generally, a matching network is used, as illustrated in Fig. 4-6.

A long-wire antenna is a good, simple method for getting moderate directivity where there is room for it. For maximum benefit, the antenna should be several wavelengths long.

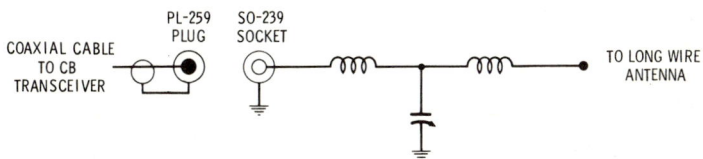

Fig. 4-6. Example of matcher for a long wire antenna.

Part 5

Feeding Antennas

Up to this point, little has been said about how the power generated by the transmitter gets to the antenna. This important matter doesn't always receive the close attention it deserves. Sometimes, tolerable results are obtained even if it is ignored. For instance, the whip antenna seen on little walkie-talkies is very like a quarter-wave Marconi. It should have a ground plane, yet reception and transmission are good enough for casual work without the ground plane. Many of the more powerful walkie-talkies are very sensibly provided with terminals for an external antenna.

A fixed CB installation usually includes an antenna a few tens of feet away from the transceiver. A special kind of wiring, called a "transmission line" runs between the two. As telegraph and Atlantic cable engineers were finding in the mid-1800s, long electric lines have odd properties. In the 1900s, the details were worked out as people learned why any line more than a tenth wavelength or so long at its operating frequency should meet some very special requirements.

47

What is a transmission line?

Physically, it is a pair of wires or a shielded cable designed for carrying radio-frequency power. The two basic transmission line types look like modified lamp cord (twin-lead), as shown in Fig. 5-1, or high-grade microphone cable (coaxial cable), as shown in Fig. 5-2, but are actually very different from these. They are constructed of far better materials, they are supplied only in certain critical sizes, and their conductor spacing, sizes, and electrical characteristics are controlled very accurately during manufacture. The best coaxial cables are

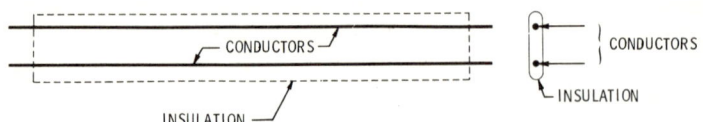

Fig. 5-1. Twin-lead antenna transmission line.

made of silver-plated copper wire and braid, but they are expensive.

The most noticeable difference of a transmission line from circuit supplies is that its manufacturer specifies the surge or characteristic "impedance" for each of his transmission lines. Typical values are 52, 72, and 95 ohms for coaxial cable, and 75 and 300 ohms for twin-lead. It is a fallacy that this has anything to do with "resistance."

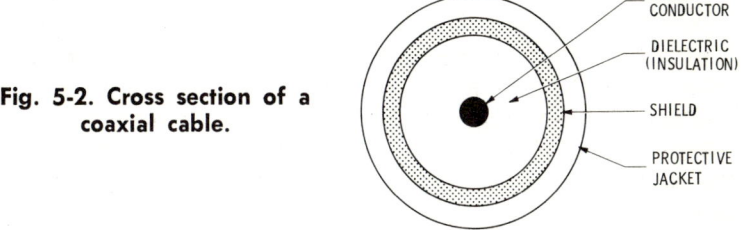

Fig. 5-2. Cross section of a coaxial cable.

The transmission line impedance is the ratio of voltage to current that flows into the line input end and within the line. It isn't resistance or "like" resistance. Thinking impedance is resistance in disguise is like expecting milk and gasoline to be alike because a quart container will hold some of each, or both. It is far better to understand the difference than learn it by experiments.

If 50 volts of rf is applied to a 50-ohm transmission line, then 1 ampere of current will flow into the line, if the line feeds a 50-ohm load, such as an antenna. Ohm's law seems to be working here, but it isn't, because the line doesn't dissipate the 50 watts flowing into it. Instead, the 50 watts flows up the line and is delivered, untouched, except for small losses caused by the dc resistance of the conductors, to the other end of the line where it is either almost all used or partially returned. This is a very different behavior than that of a resistor.

Transmission line characteristics are surprising to people with only dc circuit experience. With a little study and practical experience, understanding comes soon.

48

What is "mismatching?"

"Mismatching" is a wrong use of a transmission line. And the term "mismatching" is also much abused. Matching or mismatching are differently important at the transmission line input and output ends, and along it.

As the line works in a well-engineered installation, at any point along the line the ratio of voltage to current is its impedance. But what if the impedance changes? This happens at a very sharp bend, or a splice where the diameter ratios of inner to outer conductors are changed or their spacing is altered by mechanical forces on the line. The resulting mismatch slightly reduces line efficiency.

Matching is important at the line input and output ends. A certain ratio of voltage to current is flowing toward the output end, and it must be used up there, somehow. The element that does this is resistance. If a 50-ohm line is terminated with a 50-ohm resistor, the resistor accepts power at the same ratio of voltage to current flowing in the line. The line is said to be "matched."

But if, for instance, the 50-ohm line terminating resistor is 75 ohms, it won't accept all the power flowing in the line. Since the rule that line current and voltage must work out to its characteristic impedance *always* holds, the line adjusts this apparent impossibility by starting a reverse power flow, back toward the input. Now there are two power flows, and each meets the impedance requirement of the line. But travelling in opposite directions at the output end, they also manage to meet the 75-ohm consumption arithmetic of the resistor. The reflected power is largely lost in most systems, which is why matching at the output end of a transmission line is important.

Contrary to some belief, matching at a transmission line input end is not as important as matching at the output end. However, if a transmission line input is matched by the transmitter, maximum power will be delivered into the transmission line.

49

Why is a coaxial cable nearly always used as a transmission line?

Coaxial cable is more appropriate in CB applications than is twin-lead line. A coaxial cable is single-ended, self-shielding and insensitive to its physical environment.

Most electronic circuits are single-ended, having one signal conductor and a ground return. A double-ended circuit, by contrast, consists of two conductors at equal and opposite voltage to ground. Twin-lead is a balanced circuit. Single-ended circuits are electrically and mechanically convenient, and at least as effective as double-ended circuits. The single-ended coaxial line is naturally fed by a single-ended circuit.

Self-shielding is a bonus quality that comes with the single-ended nature of the coaxial cable. Since one conductor surrounds the other, the cable electric and magnetic fields are almost entirely within the cable. This means outside fields don't interact with the cable to induce unwanted signals and noise into the cable circuit.

Finally, also a bonus from the single-ended construction of the coaxial cable, the cable isn't sensitive to its surroundings when properly matched. It can be run through the air, along steel tower legs, through brick or reinforced concrete walls, and adjacent to other cables or wiring, all with insignificant influence upon its electrical properties, except when it is run close to high power electrical transmission circuits. This is a very useful property since coaxial cable is often installed in a pre-existing building, an automobile, or other surroundings originally designed with no thought to the needs of the cable.

50

How do coaxial cable types vary?

Coaxial cables vary in impedance, power capacity, efficiency, flexibility, size, quality, and cost. Cost is relatively unimportant where the cable is to be used for several years. Impedance is a rather critical electrical characteristic. And quality is intangible, but it has its value.

Cable impedance must meet the installation requirements, which are usually set by the antenna. Two cables of quite different impedance will look much the same. Cable impedance is determined from the manufacturer's engineering (not advertising) specifications or by measurement.

Cable power capacity and efficiency are related. In a good CB installation where some incoming signals are very weak, heavy-duty, low-loss cable is used. Heavy-duty cable, capable of handling a kilowatt or more, has lower attenuation, longer life, and greater resistance to mechanical abuse than smaller cables of similar impedance characteristics.

Flexibility and size are also related, since the smaller cable types are more flexible than the larger ones. Usually this is a matter of small importance. A coaxial cable of any size should be run in wide arcs and sweeps.

Quality and manufacturing costs are related very closely indeed. A good low-loss cable that sells on its reputation and engineering specifications may go for only slightly more than a conventional cable.

51

What kind of coaxial cable should be used?

First, the cable impedance and power rating must be appropriate for the application. Power rating is not a consideration for CB work. The necessary impedance rating is determined by the antenna, usually 50 ohms. Second, low-loss, heavy-duty cable is preferable. Its advantages outweigh its size and cost. Finally, top quality cable is indicated. Its cost is more than made up with the passage of time.

52

How should coaxial cable be spliced?

If at all possible, it shouldn't. (No spliced cable has the good physical and electrical characteristics of a continuous run.) Coaxial cable is available in runs of 100 feet or more, where a very long run is necessary. The inconvenience of managing such a long piece is offset by its reliability and good electrical qualities once it is in place.

But where a cable must be spliced, connectors are an excellent approach. Some connectors (Fig. 5-3) are the "constant-impedance" variety. That is, they are engineered to look to the coaxial cable and its signal like more coaxial cable. Assembly instructions for constant-impedance connectors may look very impressive. They aren't that bad, and once the job is done the resulting joint is easily dismantled for moving and reinstallation elsewhere.

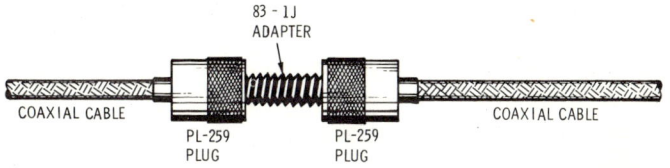

Fig. 5-3. Coaxial cable splice using uhf-type connectors.

However, a coaxial cable can be spliced like a piece of wire. It's a lot of work. The ends are opened and bared, center conductors are joined, plastic is worked in over the center conductor, and the outer conductor is joined. It's necessary to maintain impedance.

When the cable ends are joined, if the finished job is outdoors or in an unfavorable inside environment, it needs protection. A nice way to do this is to apply a couple of layers of polyethylene plastic and electrical tape overall for an airtight seal. Another approach is to apply a coat of silicone rubber, which hardens into a flexible but easily removed protective coating.

53

How are connectors attached to coaxial cable?

Specially designed connectors are required. Uhf connectors, although not constant-impedance types, are often used. A typical uhf connector assembly procedure is shown in Fig. 5-4.

54

What is a solderless coaxial cable connector?

Solderless equivalents of the popular solder-type PL-259 plug are available from several manufacturers. Firm electrical contact is made by mechanical pressure, not through solder.

55

How is a CB transceiver tuned to the antenna system?

The manufacturer's instructions should be followed. An SWR meter is usually required, connected as shown in Fig. 5-5. The transmitter output tank circuit and antenna trimmer, or pi-network trimmers, are tuned so that the SWR meter indicates maximum forward power and minimum reflected power.

56

What is a dummy antenna?

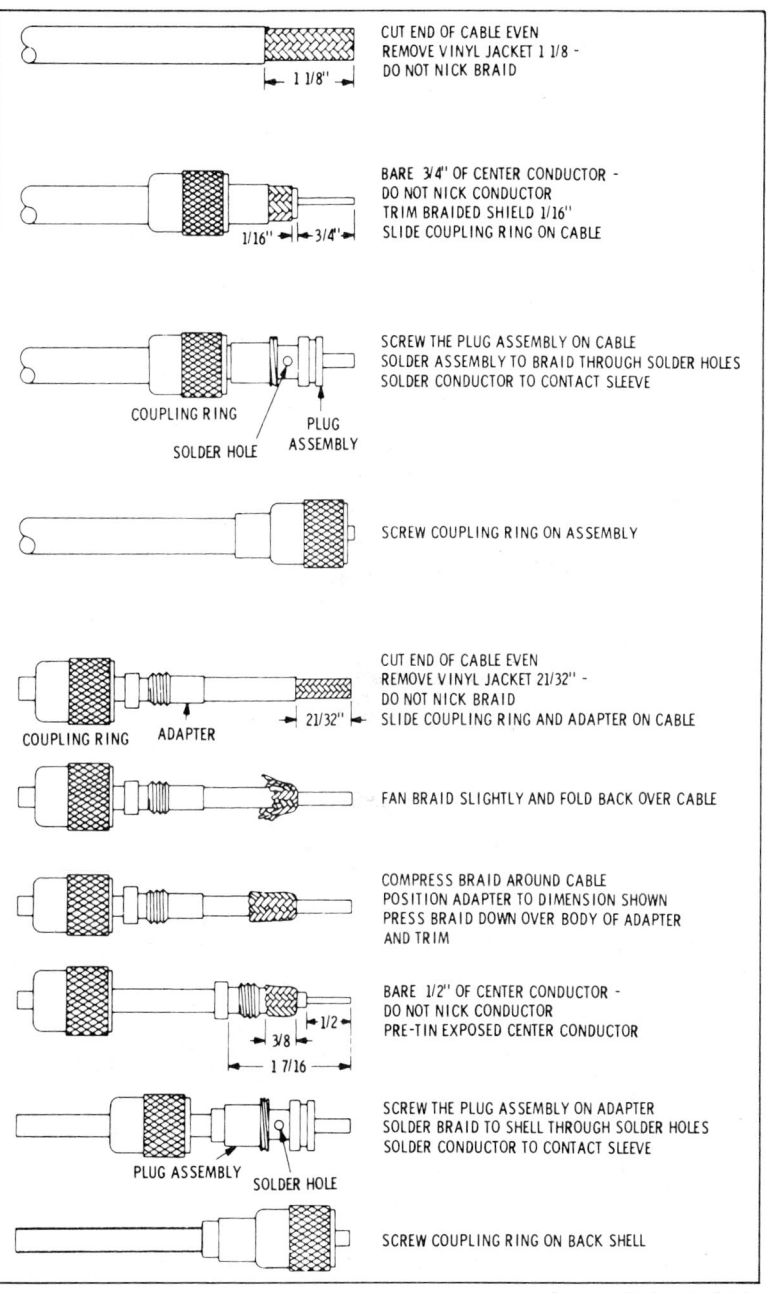

Fig. 5-4. Assembly of coaxial cable to plug.

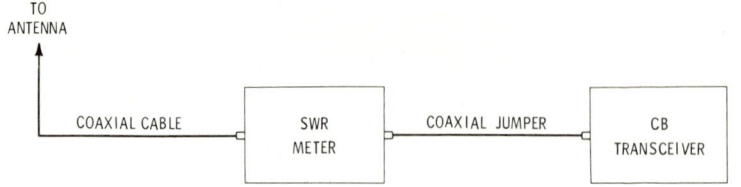

Fig. 5-5. An SWR meter is essential for correct tune-up to the antenna.

A dummy antenna is a resistive load (Fig. 5-6) with the same electrical properties as well-adjusted real antenna. It radiates no radio signal. Instead, it dissipates the transmitter output as heat. A simple dummy antenna is nothing more than a few resistors arranged to minimize inductive and capacitive reactances at the operating frequency. A more elaborate dummy antenna includes a simple rf wattmeter to indicate relative power dissipated in the resistance. A good commercial unit is calibrated directly in watts fed to it by the transmitter (Fig. 5-7).

Fig. 5-6. Example of a 50-ohm dummy load.

Courtesy Bird Electronic Corp.

A dummy antenna is necessary for the communications serviceman. Using it, he can work on a transmitter without any regard to transmission regulations or interference to outside communications traffic. He can test CB transceivers without disturbing other CB transmission. Connected to a receiver, the dummy antenna enables the technician to check the behavior of the receiver under normal operating conditions without any input signal. Even the simplest communications lab should have a dummy antenna, and can make one easily.

57

What effect does SWR have on transmission line losses?

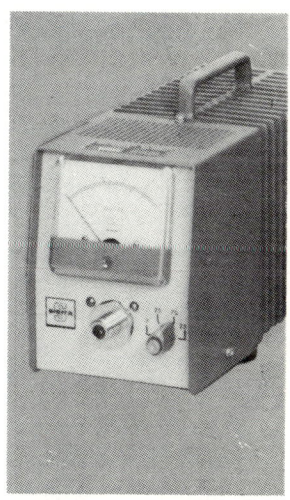

Fig. 5-7. Examples of rf wattmeters.

Transmission line losses increase as SWR (standing-wave ratio) becomes greater than 1:1. System losses increase, too. Line losses increase because the reflected power interacts with forward power to produce high-current regions in the transmission line. Power losses in these regions increase rapidly with SWR. And system losses increase, too, because power reflected from a wrongly adjusted antenna gets back to the transmitter. The reflected power is dissipated in the transmitter output stage as well as in the transmission line.

Time spent reducing SWR is well invested. It maximizes system efficiency for both transmitting and receiving, and minimizes the transmitter output stage heat dissipation.

58

How can coaxial cable be tested?

Surplus or unknown coaxial cable is risky stuff to play with. However, if it must be used, it should be tested. Here is a suggested procedure.

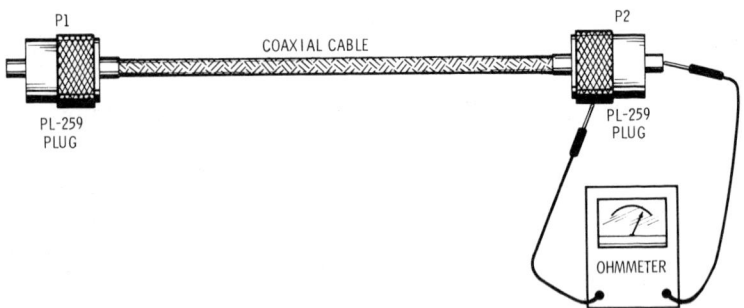

Fig. 5-8. Coaxial cable dc test. Ohmmeter should read open connected as shown, and zero resistance when P1 is shorted.

The first test is an eye check. Cracking, corrosion, mechanical damage, signs of age, and exposure to chemicals and moisture are all grounds for immediately discarding the cable. But if it looks good, maybe it is usable.

Next, dc tests establish the cable continuity and freedom from excessive leakage or short circuits. Ohmmeter tests should indicate low resistance center and outer conductors, and at least several megohms resistance between the conductors (Fig. 5-8).

To measure transmission loss, connect an rf wattmeter (Fig. 5-7) to the transceiver through a coaxial cable jumper and note measured power output, as shown in Fig. 5-9A. Then connect the cable to the transceiver and the rf wattmeter to

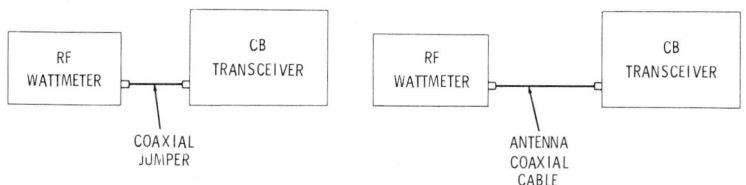

(A) Transmitter output. (B) Transmitter output minus coaxial cable loss.

Fig. 5-9. Coaxial-cable transmission loss measurements.

the far end of the cable, as shown in Fig. 5-9B, and again note measured rf power. Then refer to Table 3-1 to determine loss in dB. If, for example, the power level at the far end of the cable is one-half that at the transmitter output, the transmission loss is 3 dB. Table 5-1 lists the rated losses of standard types of coaxial cable.

Table 5-1. Coaxial Cable Transmission Losses at 27 MHz

Cable type	Loss in dB		
	50 feet	100 feet	200 feet
RG-58/U	1.0	2.0	4.0
RG-8/U	0.5	1.0	2.0
Andrew FH-1 (¼")	0.43	0.86	1.72
Andrew FH-2 (⅜")	0.3	0.6	1.2
RG-17/U	0.25	0.5	1.0
Foamflex (½")	0.25	0.5	1.0
Spirafil (½")	0.2	0.4	0.8
Spir-O-Line (½")	0.2	0.4	0.8
Andrew FH-4 (½")	0.18	0.36	0.72
Andrew FH-5 (⅞")	0.12	0.24	0.48
Styroflex (⅞")	0.1	0.2	0.4
Spir-O-Line (⅞")	0.1	0.2	0.4

Part 6

Installing Base-Station Antennas

Of course you can run out for an antenna, buy the most attractive one in sight (probably at the highest price), and nail it up in the air. This book is meant to encourage a more leisurely, thoughtful pace that will get you far better results.

Good procedure is to examine the tradeoff situation first. Write down what's important. What should the antenna achieve? What minimum antenna will do a good enough job? Can you test your ideas somehow? Choose a radiation pattern and an antenna gain; consider prices versus benefits and service and adjustment convenience. Study—the knowledge has its own value and may some day be worth a better job to you. If your judgment lacks experience, call in friends. At last, choose some sensible all-around compromise, and proceed with the installation.

Could this come out wrong? Yes. Could it be omitted, and yet you'd put up a highly effective antenna? Perhaps you know somebody who says he did, and perhaps he doesn't know the difference.

59

What is the difference between antenna height and effective antenna elevation?

Antenna height is measured from the ground or from the top of the permanent structure it is mounted on. In CB radio applications, "antenna height" has a definite legal meaning and must not exceed 20 feet.

Antenna elevation is something else. If you happen to live on a 500-foot local bump in the ground, the FCC doesn't expect you to dig a 480-foot hole for your antenna. The legal requirement of 20 feet starts at the top of a hill, trees, a building, or a natural feature the antenna is mounted on (Fig. 3-2).

60

How is the legal 20-foot height limit applied?

The Part 95 regulations, Section 37, state four general restrictions in mounting a CB antenna. First, the antenna height cannot exceed one foot per 200 feet from the nearest airport boundary. Second, the antenna height cannot exceed 20 feet above the ground, natural formation, tree, or building it is mounted on. Third, the antenna can be mounted at any height on the tower of an existing separately licensed transmitting station tower, but must not extend above it. Last, the antenna can be mounted on any receiving antenna tower that meets the airport and CB antenna height restrictions (Fig. 6-1).

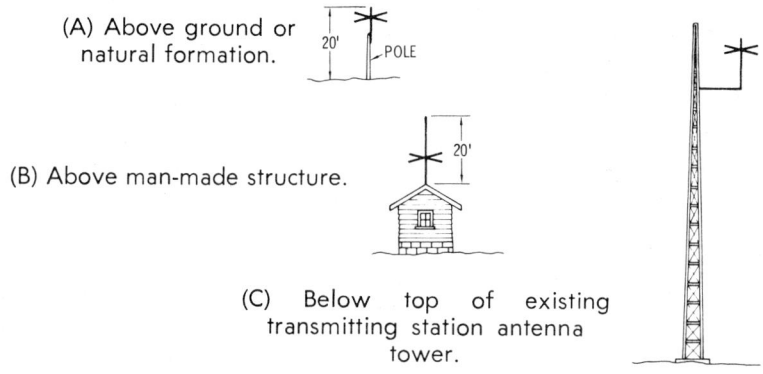

Fig. 6-1. CB antenna-height limits.

61

How can TVI be eliminated or reduced?

A shotgun approach is very ineffective against TVI (television interference) problems. It is important to understand how TVI arises, and what steps work against the three basic kinds of cases. These are TV receiver overloading by the transmitter signal, harmonic radiation from the transmitter, and intermodulation between the transmitter and some other signal at a point not very distant from the TV antenna.

If the transmitter signal enters the TV receiver too strongly, either by way of the TV power cord, from the TV antenna, or through direct pick-up by an inadequately shielded TV receiver, the TV receiver input circuit will become overloaded. Since a TV receiver is a relatively complex system, symptoms can be generated ranging from apparent loss of signal to a variety of odd picture shapes and sound outputs. The transmitted CB signal might be heard through the TV set loudspeaker.

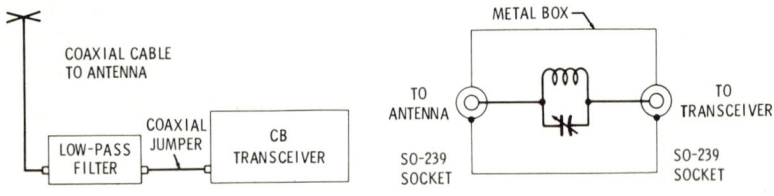

(A) Low-pass filter at transmitter.

(B) Parallel-resonant wave trap tuned to second harmonic (approximately 54 MHz).

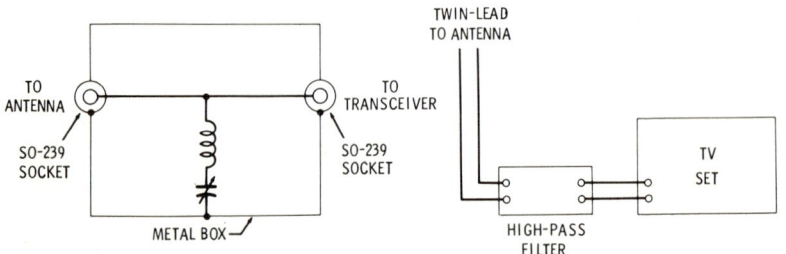

(C) Series-resonant wave trap tuned to second harmonic (approximately 54 MHz).

(D) Filter at TV receiver.

Fig. 6-2. TVI suppression techniques.

Overloading remedies are a line cord filter, a high-pass filter at the TV set antenna terminals to reduce the level of the signals reaching the receiver, and relocating the TV or CB antenna so that less CB signal reaches the TV set.

CB transmitter harmonic radiation is simply the transmission of multiples of the transmitter operating frequency. A poorly designed transmitter or one that is wrongly adjusted (and probably inefficient because of this) are common reasons for harmonic radiation. Readjusting the transmitter may solve the problem. It is also possible to install a low-pass filter in the antenna transmission line of the CB set to reduce the harmonic radiation. A simple wave trap circuit added to the transceiver may also do the job. Fig. 6-2 illustrates various TVI suppression techniques.

Finally, intermodulation from some outside source can be a very tough problem. A variety of cruddy metallic joints can serve as radio detectors. Such joints occur in guy lines, tin roofs, rainspouts, and elsewhere around a house or car. Where such a joint carries both CB and other transmitter signals, it mixes the two signals to provide a variety of new ones—all bad. And such a joint can look just like twenty good ones nearby. Once it is found, a good hammer blow or some cleaning and soldering will fix the problem permanently.

62

What is an antenna matcher?

An antenna matcher is a special variable circuit connected between the transceiver and the antenna to improve the system efficiency. There are many kinds of antenna matchers. Some are provided with instruments measuring rf voltage or current, or with forward/reflected power meters. These indicate how effectively the system is working. Other matchers make do with existing instruments such as a built-in transceiver power-output meter (Fig. 6-3).

Some matchers are designed to couple rf power into almost anything from the proverbial wet clothesline (which has been seen to work, using salt water) to the kitchen screen door, a water tower, or even a real antenna. Another class of matchers is designed to couple rf power into a specific kind of antenna such as a long-wire, an end-fed Marconi, or various tuned-line balanced antennas.

63

Is a lightning arrestor required?

A lightning arrestor or its equivalent, such as safety ground switch, does nothing to improve CB communication, but is extremely important in areas which experience thunderstorms. The antenna system without an appropriate safety arrangement is dangerous to the operator, the equipment, and the

Courtesy Lafayette Radio Electronics Corp.

Fig. 6-3. Meter indicates relative rf power output and incoming signal strength.

general property. A fire insurance policy may become void without notice if an unprotected or wrongly protected antenna is installed on the premises. The proposed antenna installation should be discussed with the fire insurance people before its completion.

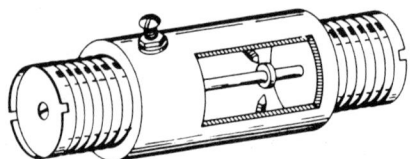

Fig. 6-4. Lightning arrester used in CB installations.

The most common type of lightning arrestor used at CB base stations is a cylindrical device inserted between the antenna transmission line and the transceiver antenna receptacle (Fig. 6-4). Its shell has a terminal which is connected to an earth ground.

64

What are some basic precautions that can be taken against lightning damage?

Lightning can strike almost anywhere. There are at least a few thunderstorms every year in all parts of the country. The insurance people are well aware of this, and write policies that are self-canceling if the house owner increases the hazard to his property, such as a wrongly installed antenna system.

Lightning is simple and easily understood. It's the fireworks that make it impressive, and the fact that when it strikes there's so much of it so suddenly. A strike is the end of a series of events. An electric charge in the clouds induces a "bound charge" in the earth. A discharge follows when the voltage between a cloud and an earth-bound charge is great enough. The cloud charge is uncontrollable, but the earth-bound charge can be leaked off harmlessly through points on the antenna, the tower, or the lightning system. Because of this leakage, protected buildings are not only less likely to be harmed by a strike, but they are struck less often.

True "lightning arrestor" arrangements come in where a strike actually occurs. There isn't any such thing as a "lightning-stopper" but the arrestor system can shunt the thousands or ten-thousands of amperes harmlessly to ground. The basic arrestor system for CB is the self-acting spark gap in the coaxial cable circuit.

65

How should a CB antenna be mounted with respect to a TV antenna?

It should be mounted horizontally some distance from a TV antenna, and above or below it. If it is too near, a significant amount of power from the CB transmitter may reach the TV set input circuit, degrading reception.

66

Can a CB antenna be used with both a CB transceiver and a 30-50 MHz band monitor receiver?

Yes. A device can be used which allows signals to be fed simultaneously to both the CB set and the monitor receiver. A typical circuit is given in Fig. 6-5. When the CB set is transmitting, the tube of the device (VIA) cuts off and protects the monitor receiver from damage by the CB transmitter signals.

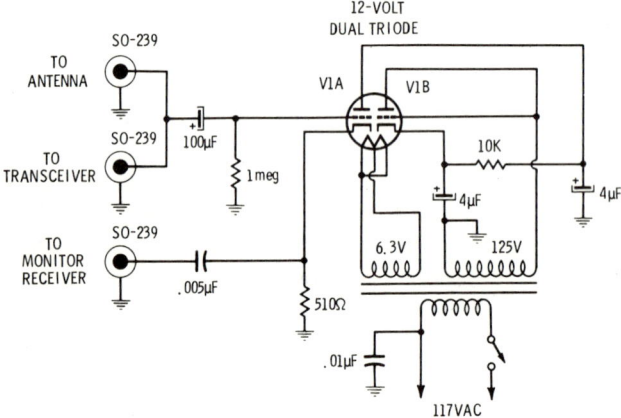

Fig. 6-5. CB transceiver/monitor receiver antenna coupler.

67

Can a TV antenna rotator be used with a beam-type CB antenna?

Of course, if the CB antenna is not too large or too heavy.

68

What kind of antenna can be used on an apartment terrace?

Two important factors in apartment terrace antenna mounting are antenna visibility and antenna grounding. Some CB antennas are extremely visible and may provoke unreasoned and unpleasant reactions from nearby residents. An approach sometimes used by careful radio amateurs is to erect some unusual antenna, and then leave it unconnected. Each neighbor who brings in some complaint of severe interference, etc., is shown politely that the antenna isn't connected yet—the cable is back-ordered and won't be in for some time. When this static dies down, the installation is unobtrusively finished. A few people will attribute very remarkable effects to antennas, even unfinished ones.

Where visibility is a factor, a small mobile-type antenna can be grounded to the building frame, or installed in an electrical mockup of the mobile installation it was designed for. A converted beach table steel top would make a workable short-antenna ground plane, and with a wooden umbrella mount could do double duty. A variety of arrangements are possible using wire mesh or screen.

But if grounding is really a problem, then it is necessary to use an antenna that requires no ground: marine-type CB antenna, for example. A dipole or a loaded, shortened dipole can also work out very well.

69

Can a CB antenna be mounted in an attic?

An attic with adequate headroom is an excellent antenna site, if the roof over it is free of metal objects larger than nails. But sheet metal in the roof angles or a metal roof are very unsatisfactory, because this will degrade antenna performance for any attic mounting.

70

How can a CB antenna be mounted on a water tower?

Since the tower is a permanent structure, Part 95 regulations permit the antenna to extend 20 feet above it. There is another legal matter. The tower owner can request a written agreement concerning the use of his property. The matter should be settled before any money and much time go into the project. Mechanically, the installation must be safe to operator, installer, and passers-by. The tower is likely to be a very rugged structure, so this matter is entirely in the hands of the CBer.

Two matters come up from the electrical view. The first is lightning. The tower elevation increases the danger, and the antenna on its top gets this benefit plus a bonus. A really good ground connection is indicated, particularly if the tower is of wood construction or has a wooden tank.

And the other electrical matter is radio performance, the reason for putting the antenna up there in the first place. The antenna may be clear of the tower and depend entirely upon its own design for its characteristics, or the tower can serve as part of the antenna. A large metal surface could serve as a ground plane for a mobile whip-type antenna.

71

How can coaxial cable be run from outside into a building?

Special feedthrough connectors are available for feeding coaxial cables through walls. A piece or two of wood with appropriate notches or holes may be placed in a window. The window is pulled down on it and sealed at the top with pieces of polyurethane foam. An unobtrusive hole may be drilled in a wall, taking notes from telephone company practice. Whatever arrangement is used, adequate lightning protection is a must.

72

How can electrical connections be kept from corroding?

The "corrosion" problem faced by electrical connectors in the open air ranges from tarnishing in a very dry area to literal dissolution in damp salt-water areas. The remedies are to use good connectors to begin with (some are silver-plated pot metal, which is unsatisfactory) and to keep them dry and free of chemicals.

Various schemes for protection range from mechanical protection in a can with a rag stuffed in, or a plastic bag open at both ends and sealed with tape after the connection is made, to covering the connection with a silicone plastic. The bag scheme is effective because the protected joint is easily checked. The silicone scheme is mechanically good because the plastic is flexible and inexpensive. Silicones are available in hardware stores and boating shops. Epoxy resins turn hard and brittle and are not suitable for protecting coaxial cable joints.

73

Why should a tower be accurately vertical?

A tower should be accurately vertical to minimize "bending moments." To illustrate bending moments, grasp a small stick by the ends and bend it. The load it can carry in this matter is much less than the compressive load it can carry while perfectly straight. A similar situation appears in towers.

If the tower is truly vertical, the basic antenna load and the tower weight are almost perfectly compressive. The tower is very strong against these loads. It is able most effectively to support additional loads caused by wind, icing, or somebody climbing on it. But if the tower is bent or tilted, then its normal loads develop as additional bending moment and the tower is less able to survive added load.

Also, a bent or tilted tower looks ugly and disreputable. It isn't hard to get a tower truly vertical. A six-inch or one-foot carpenter's level works well for short, stiff towers. It is used to square the tower north-south, and then east-west. This squares the tower from all angles. For taller towers, a plumb line some distance away serves as a reference for guy line adjustments. Again, the tower is squared north-south, and then east-west.

74

How should a tower be guyed?

A tower should be guyed generously; the details depend upon the tower stiffness. Some towers are so stiff that they get by with little or no guying. Guying isn't expensive unless the top load is very great.

Nylon line is very strong and has a desirable resilience. There should be provision for controlling its tension. Steel wire is also good, but it needs closer attention. Turnbuckles, when used, should be oiled once in a while so they do not rust fast. Safety wiring may be in order, too. A four-wire guying system is easier to install and adjust than a three-wire system.

Part 7

Installing Antennas on Motor Vehicles and Boats

Practical limits of vehicle size, antenna size, and mechanics make mobile installation easier than fixed installation work. But beyond that, it's all uphill.

The restricted size of a mobile antenna reduces radiation resistance, which works strongly against overall efficiency. An earth ground or a large ground plane cannot be used to increase effective antenna size. An apparently nondirectional antenna shows better performance diagonally forward over a car than in any other direction, if it is mounted on the rear bumper. Where the principles described in this book seem to have been successfully ignored—they haven't. A performance price is paid somewhere.

There are some very ingenious ideas for installing mobile antennas. These are touched on here. For more facts on the rapidly changing mobile-antenna field, refer to recent catalogs.

75

What is the best place on a car for mounting an antenna?

The best place on a car for mounting an antenna is in the center of the roof. This site gives the most symmetrical current distribution and the most equal arrangement of the electric lines of force of the antenna. If the antenna is moved along the car roof toward the back, its pattern will improve toward the front of the car. The antenna acts as if it were installed on a ground plane with a deep notch; its performance is poorest in the direction of the notch. Wherever an antenna is mounted on a car bumper, it will show an unavoidable directional characteristic (Fig. 7-1).

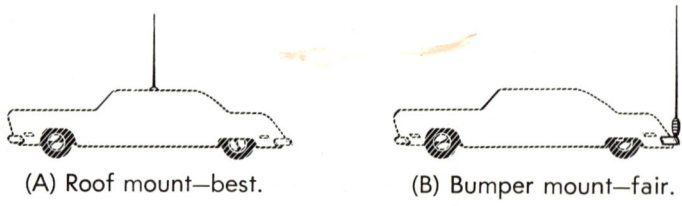

(A) Roof mount—best. (B) Bumper mount—fair.

Fig. 7-1. Typical mobile antenna installations.

76

What mistake in a truck or van antenna installation leads to poor results?

A bumper-mounted whip on the side of a van or truck body, as shown in Fig. 7-2, can be expected to work poorly. Its radiation pattern will be near zero on the side toward the body, and poor in other directions. Since the sheet metal body prevents good antenna coupling into space, the antenna will have a very low radiation resistance and be hard to feed. A smaller antenna, mounted on the vehicle top, will usually work much better.

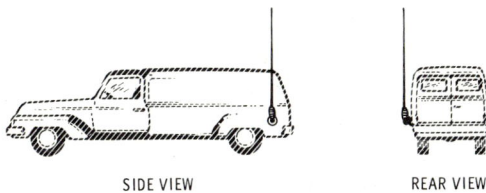

SIDE VIEW   REAR VIEW

Fig. 7-2. Unsatisfactory antenna installation on a truck.

Can the same antenna be used for CB and an auto radio?

Yes. Special coupling devices are available that enable a car radio and a CB rig to use the same antenna. The antenna is a whip externally identical to an ordinary car radio antenna. An installation like this has great value in a large city. It most definitely is no gimmick. CB rigs and other electronic equipment interests thieves; a concealed rig using an apparently conventional antenna is excellent practice in bad areas. Plastering a CB call number all over some window is a giveaway, too.

78

Where is the loading coil located in an electrically shortened antenna?

The loading coil is located at the bottom or far up the antenna. Bottom-loading is often used. But for maximum efficiency, the coil should be part way up the antenna to expose the maximum current-carrying conductor to space. Many commercial mobile antennas are "center-loaded" (Fig. 7-3). The coil is not placed at the top because the current there is near zero,

Fig. 7-3. Center-loaded mobile antenna.

and the coil would have little effect. But if the coil is midway along the antenna there is an exposed lower conductor carrying a high current and an exposed upper conductor for current to flow into.

79

What is capacitive loading?

Capacitive loading is the addition of a nonradiating section at the top of a shortened antenna to lower its resonant frequency and emphasize its current-carrying portion. Used

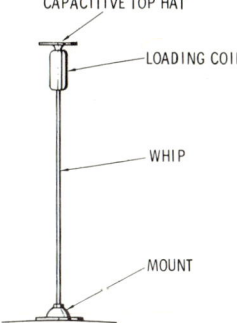

Fig. 7-4. Top-hat CB antenna.

with inductive loading at the top end of its vertical portion, a capacitively loaded antenna offers the maximum effectiveness possible from a simple structure (Fig. 7-4).

80

What are the sources of objectionable mobile noise?

Impulse noise from the operator's car and from other cars on the highway is often severe. Nothing can be done about the other cars, except by the government. The operator's car can be equipped with ignition noise suppressors or with a resistance type high-tension wiring to reduce noise generation. Ignition noise may also be coupled into the transceiver through the car wiring. This is reduced by bypassing the ignition system supply lead near the coil.

In the process of generating electricity, sparking at the generator commutator or slip rings can cause interference. After ensuring that these rotating contacts are in good condition, the noise can be further reduced by installing bypass capacitors at the generator terminals. The voltage regulator may require attention too.

The car in motion, and its wheels and tires, are also sources of electrical noise. This originates in the friction of the car against the road and the motion of the car through the air. The insulating grease in the wheel bearings then breaks down periodically, producing an irregular noise with great nuisance potential. Special springs and brushes are available for making a good electrical connection across front and rear wheel bearings. These must be replaced periodically.

The motion through the air can generate enough electricity to cause corona discharges from sharp points. The mobile antenna is usually the sharpest point. This corona noise can be reduced or eliminated by mounting a small metal sphere on the antenna end, or insulating the antenna with a length of electrical tape.

81

How can an external antenna be connected to a walkie-talkie?

Many walkie-talkies are provided with a jack for connecting an external antenna to increase transmitting and receiving effectiveness. One can be added as shown in Fig. 7-5. An antenna connected to this jack works like any fixed-base station antenna. Since the walkie-talkie integral antenna cannot be really effective, the range and reliability of the unit will be greatly improved by even the simplest external antenna.

Fig. 7-5. Circuit for adding an antenna jack to a walkie-talkie.

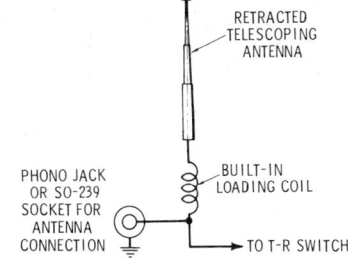

82

Is it necessary to ground an antenna to a car body?

If the antenna is basically a quarter-wave Marconi, its coaxial cable outer conductor must be grounded to the car body. A marine-type antenna of the nongrounded variety might not require any connection to the car body.

83

What kind of antenna can be used on a convertible car?

A bumper or cowl mounted quarter-wave or loaded whip is the best bet. Unfortunately, a metal top is not available for use as a ground plane.

84

What is a magnetic mount antenna?

A rented car, or one borrowed from a friend, isn't any candidate for punching holes and making permanent antenna installations. Some manufacturers provide special antennas that are held in place by a strong magnet and a friction surface.

Also, antennas with a built-in magnetic mount are available. The antenna is placed on the car roof or trunk lid, and the cable is brought in through the window.

85

What is a clip-on antenna?

Several manufacturers offer loaded antennas which clip onto a car rain gutter (Fig. 7-6). They work all right if the clip makes good electrical contact with the car body.

Fig. 7-6. A gutter-mount mobile antenna.

86

What is a side-mount antenna?

Special antennas are available which can be attached to the side of the antenna tower or building (Fig. 7-7). Generally, they will not provide omnidirectional coverage because of the effect of the mounting structure.

Fig. 7-7. A side-mount antenna.

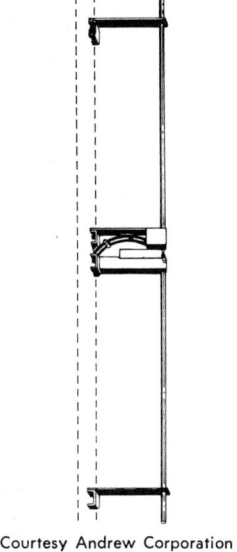

Courtesy Andrew Corporation

87

What is a spring mount?

A spring mount is almost obsolete, but not quite. When most mobile CB antennas were quarter-wave whips (100-110 inches), a spring mount was almost essential. A spring mount (Fig. 7-8) protects the antenna when it is struck because of low overhead clearances.

88

How can an antenna be mounted on a car without drilling holes?

No-hole installation systems include magnetic-mount, clip-on, bumper mount, and adhesive installation. However, for maximum reliability and performance, an antenna should be mounted permanently. It costs very little to have the hole patched up when trading in a car. And, in most cases, it is extremely important that the antenna base make secure electrical contact with the car body.

Fig. 7-8. Spring-mount assembly for bumper.

Courtesy Master Mobile Mounts

89

Is there such a thing as a mobile beam antenna?

At least two manufacturers offer mobile beam antenna kits which improve communication in two opposing directions. They consist of two antennas spaced by a critical distance and a phasing network. For certain applications, they're great, but they're not for general purpose use except when means are provided for cutting out one antenna and using the other for omnidirectional transmission.

90

What kind of antenna should be used on a boat?

A wooden boat in fresh water is no place for any antenna that requires a ground connection. There is a very simple solution to this engineering problem. A dipole, since it is balanced, doesn't need any ground, and can be used on a wooden boat.

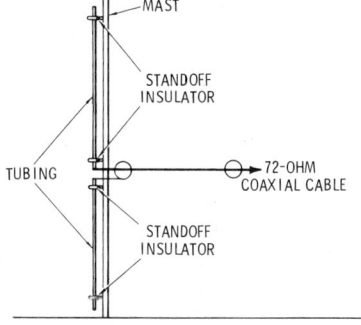

Fig. 7-9. Home-made vertical antenna for a boat.

A no-ground dipole boat antenna can consist of two 9-foot lengths of copper or brass tubing, vertically mounted. The 72-ohm coaxial cable is brought out horizontally from the center point of the dipole, as shown in Fig. 7-9. But a factory-made marine-type CB antenna (half-wave or five-eighths-wave) which requires no ground plane or hull ground is better.

91

Can a wire antenna be installed on a boat?

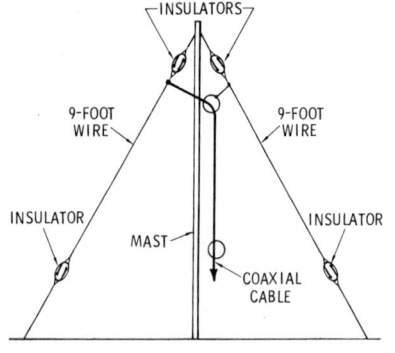

Fig. 7-10. V-shaped dipole on a boat.

Of course. A V-shaped dipole can be attached to the mast, as shown in Fig. 7-10. Or, a single wire, 18 feet or longer, can be fed through an antenna matcher, as shown in Fig. 7-11.

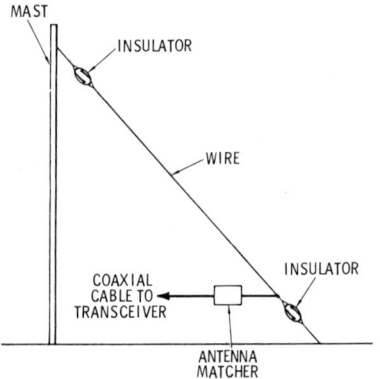

Fig. 7-11. Long-wire antenna on a boat.

Part 8

Testing Antennas

The careful worker feels a certain cautious suspicion when his antenna installation is finished. He knows he has planned well. He also knows he cannot see or feel what is happening up there, and things could be well off optimum with no clear indication of trouble. Again, he may be looking at an antenna whose performance is suspected, but not known, to be poor. Just looking at the installation helps, but it's not enough.

Only test instruments can supply facts. A report from any other source only says, "It seems to work." This isn't definite enough for new work, and the operator may reply about an older installation, "But I think it worked better a couple months ago."

The basic test instruments are a field strength meter, a directional power (SWR) meter, and good judgment. The field strength meter provides a measure of system performance, the power meter tells something about events inside the system, and the seasoned judgment of a knowledgeable technician assembles these results into sound conclusions.

92

What preliminary check may make electrical tests unnecessary?

A preliminary check is a good close look at the installation. The antenna system may exhibit serious design or installation defects. Facts obtained during the inspection will make test results more meaningful. Maybe the sight check will disclose there just isn't any percentage in testing—that good new work is needed first.

A good sight check includes the immediate environment, and farther out. Poor communication may follow the recent installation of new local power wiring. A rising steel-framed building nearby may be shielding a large piece of countryside that formerly was well within reach. Once these details are settled there is plenty of time for the tests, and the results will mean more.

Any antenna system should be checked at least once a year.

Points to look for are parts wear, cable deterioration, rust, corrosion, and a gradually developing unsafe construction.

93

Why is a CB receiver a poor test instrument?

First, a CB receiver is a poor test instrument because the natural responses of the ear minimize the effects of signal strength changes. A doubling or halving of signal strength is hardly noticeable. It makes some sense if such changes are hardly noticeable that they are hardly important. In practice, things are more complicated, and smaller signal level changes other than doubling or halving are technically significant.

Second, all modern radio receivers include an avc or agc circuit. Either circuit continuously adjusts receiver sensitivity to minimize the audio output level change for a tremendous range of signal input levels. A good receiver won't show noticeably different responses to two signals when one is 10,000 times stronger than the other.

If the receiver has an "S" meter, its value for test work increases. The "S" meter indicates, very roughly, how much the agc or avc circuit is compensating a strong or weak signal, and so it carries news about signal strength. This result is by no means a laboratory result, and two specimens of the same model receiver are likely to yield similar results. Also, there is no agreement among manufacturers regarding what signal input levels will produce given "S" meter readings.

94

How can antenna radiation be monitored?

A whip antenna is placed near the antenna to pick up a little radiation. It must be far enough from the antenna to detect true radiated power, not the currents circulating in the antenna structure. A few times the maximum length of the antenna will be far enough.

The rf picked up by the whip is rectified and fed to a meter. Meter response is calibrated to full scale under best possible conditions, and if the antenna is rotatable, the meter readings are recorded for various antenna headings. This scheme

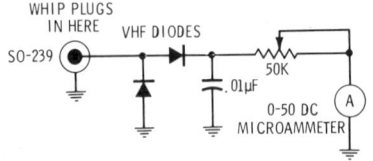

Fig. 8-1. Simple field-strength meter.

catches system degradation early and can be used to compare the performance of various transmitters or control settings. The schematic in Fig. 8-1 shows an untuned, sensitive indicator. It will not generate TVI or change CB antenna performance.

95

What is a comparison test?

A comparison test is a test in which circuit or gear performance is compared with some standard or other circuit, rather than being measured. Feeding two transmitters into a dummy load with a simple rf voltmeter is a simple comparison test that tells which works best. Checking receiver "S" meter readings from the same remote transmitter at about the same time, using two different receiving antennas, is a reliable way to find out which antenna picks up the most transmitted signal.

Comparison tests can even give accurate results. For instance, the actual gain of a gain antenna can be measured by comparison testing. This requires a receiver with an "S" meter, a resistive attenuator, a good dipole, the gain antenna, and a fixed transmitter system at a reasonable distance.

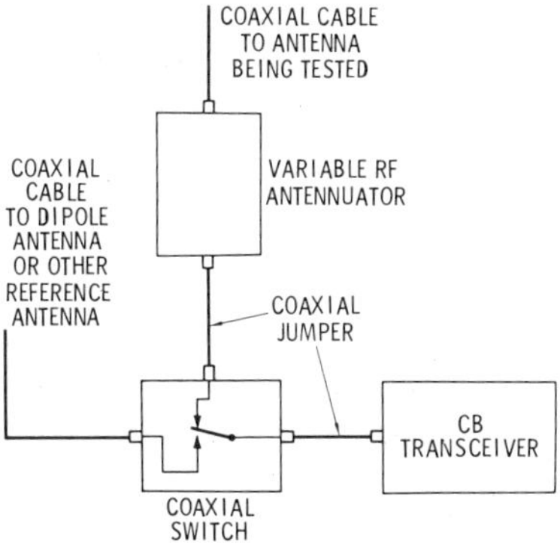

Fig. 8-2. Setup for antenna comparison test.

```
                    COAXIAL CABLE
                    TO ANTENNA
                    BEING TESTED
                         |
        COAXIAL      ┌────┴────┐    VARIABLE RF
        CABLE        │         │    ANTENNUATOR
        TO DIPOLE    │         │
        ANTENNA      │         │
        OR OTHER     │         │──── COAXIAL
        REFERENCE    │         │     JUMPER
        ANTENNA      │         │
           │         │         │
           └─────────┤         ├──── CB
                     │         │     TRANSCEIVER
                     └─────────┘
                      COAXIAL
                      SWITCH
```

The dipole antenna is set up, and the receiver "S" meter reading is noted. Changing nothing else, the dipole is replaced by the gain antenna, and the attenuator (Fig. 8-2) is added to the transmission line. Then the signal attenuation required to get the same receiver performance as before is noted. A signal that requires 10 dB of paring to reach the original reference or dipole level must have been 10 dB larger to begin with. Since only the antenna is changed, the new antenna gain must be 10 dB over that of the dipole.

96

What is an attenuator?

An attenuator is a resistive network for reducing an audio or radio signal. The receiver gain or volume control is a simple variable attenuator circuit. The attenuators used in radio frequency transmission lines for signal reduction and comparison tests are more elaborate.

Such attenuators must look to the transmission line like more transmission line. This prevents setting up unwanted signal reflections, or if such reflections exist it does not upset them by adding new ones. For instance, if an attenuator is connected into a 50-ohm coaxial line it must look like more coaxial line from either end and at any setting. This is fairly easy to do at CB frequencies, and the CBer who cannot afford to buy a commercial attenuator can make a very satisfactory one from composition resistors and inexpensive slide switches (Fig. 8-3).

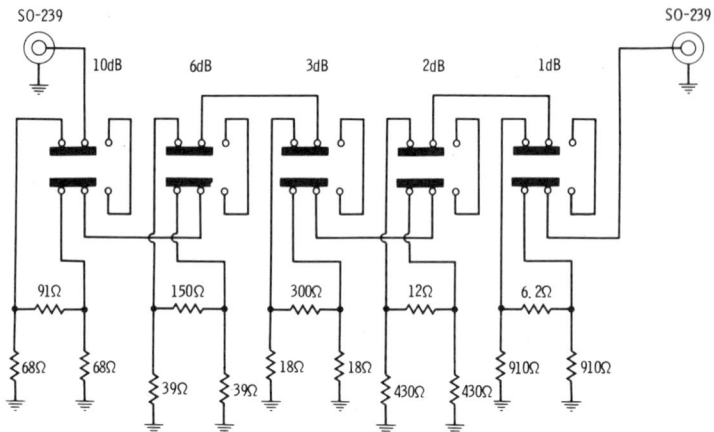

Fig. 8-3. Attenuator with 22-dB range in 1-dB steps.

97

Why is reflected power important?

Reflected power means that something is wrong. If the system is designed and working ideally, power flows from the transmitter through the transmission line into the antenna and out to space without any reflection anywhere.

Power is reflected by any of three conditions. More than one may appear in a particular case. The first is a resistive mismatch, ranging from a wrong resistive load to an open or a short circuit. The second is a capacitive component in the load. This stores power for part of the rf signal cycle and then returns it to the transmission line. The third is an inductive load component, which also stores and returns power. It often happens that power is reflected by a resistive plus a net capacitive or inductive—usually termed "reactive"—mismatch. Tests measuring amount and kind of reactive mismatch are fairly complex.

Good installation design minimizes reflected power by using engineering specs for antenna and transmission line to guarantee correct matching upon assembly. Then, during initial performance checks, the antenna is "tuned." That is, it is adjusted so that its inductive and capacitive reactances balance at the operating frequency. The result is almost zero reflected power.

Even when the antenna is correctly tuned, its forward and reflected power should be checked. Failing forward power is an early sign of transmitter trouble, and increasing reflected power reliably indicates a need for antenna maintenance.

98

How is reflected power measured?

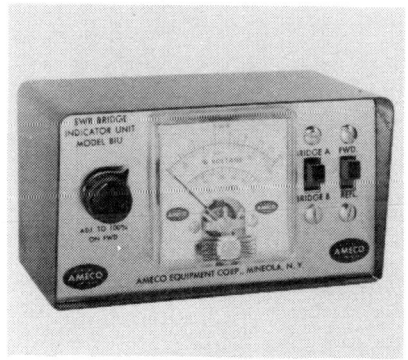 

(A) Unit by
Ameco Equipment Corp.

(B) Unit by
Seco Electronics Inc.

Fig. 8-4. Examples of SWR meters.

A so-called "SWR" meter is used for measuring reflected power. The meter circuit is a simple arrangement of two tiny loops connected at one point to the center of a short transmission line length. These loops run very near the center conductor for a little distance, but in opposite directions. Their combined inductive and capacitive coupling to the cable results in direction sensitivity; one loop responds only to power travelling up the line, and the other to returning power.

Operating technique is to adjust the transmitter circuit for maximum scale reading (power travelling toward the antenna, or "forward"). Then the meter circuit is switched to the "reverse" pickup loop, and the resulting reading is compared with the forward value. SWR meters are calibrated in terms of forward and reflected power indicated on a meter, and are connected in the circuit as previously shown in Fig. 5-5. Fig. 8-4 illustrates a typical SWR meter.

99

How is "VSWR" misleading?

This term is often applied to a convenient test unit that doesn't measure VSWR. VSWR, or voltage standing-wave ratio, is simply the ratio of maximum to minimum voltage appearing along a several-wavelengths long transmission line. It's necessary to test along a long line to measure it, making several voltage tests.

The typical "VSWR" meter doesn't measure voltage at all. Its pickup system is sensitive to directional power, up the line, or down it. Then, the meter calibrations tell what the manufacturer thinks is the VSWR for the observed power flows. This is an interpretation, although a useful one. The VSWR meter is often and more correctly called a "Forward/Reflected Power Meter."

100

What can be learned by measurements at the antenna terminals?

Antenna terminal measurements are useful because the antenna is actually the most uncertain part of the installation.

But since antenna characteristics are affected by nearby objects, the tests must be done carefully. Measurements at the terminals of the antenna provide facts concerning its resonant frequency, its input resistance, its input reactance dependence on frequency, and its bandwidth.

Resonant frequency can be tested by adding a hairpin loop across its terminals and checking with a grid dip meter. This makes rough tuning possible, minimizing adjustments made when the antenna is relocated in its operating position.

Input resistance varies remarkably as the antenna is moved away from the earth. Dipole input or radiation resistance varies with antenna height in wavelengths. Many simple instruments for checking antenna input resistance are described in radio amateur magazines such as *QST*, *73*, and *Ham Radio*.

Antenna reactance can also be measured at its terminals, but the test is a rather difficult one, and it is not usually important for CB work. As a rule, the antenna is resonated, and its reactance is assumed unimportant. Antenna bandwidth can also be measured at its terminals, but this test is rarely done.

101

What transmission line lengths are preferable for test work?

If the transmission line is any number of electrical half waves long, electrical conditions at the antenna terminals are reproduced at the input end of the line. It is important to use good coaxial cable if the antenna is to be tested through it, because line losses will otherwise make the antenna look better than it really is. A multiple half-wave transmission line allows the operator to check the antenna from the operating position.